BIOLOGY

The Unity and Diversity of Life

STARR

TAGGART

EVERS

STARR

Volume 2
Evolution of Life

14TH EDITION

Australia • Brazil • Mexico • Singapore • United Kingdom • United States

Evolution of Life
Biology: The Unity and Diversity of Life,
Fourteenth Edition
Cecie Starr, Ralph Taggart, Christine Evers,
Lisa Starr

Product Director: Mary Finch

Senior Product Team Manager: Yolanda Cossio

Senior Product Manager: Peggy Williams

Associate Content Developers: Kellie Petruzzelli,
Casey Lozier

Product Assistant: Victor Luu

Media Developer: Lauren Oliveira

Senior Market Development Manager:
Tom Ziolkowski

Content Project Manager: Harold Humphrey

Senior Art Directors: John Walker, Bethany Casey

Manufacturing Planner: Karen Hunt

Production Service: Grace Davidson & Associates

Photo Researcher: Cheryl DuBois, PreMedia Global

Text Researcher: Kristine Janssens,
PreMedia Global

Copy Editor: Anita Wagner Heuftle

Illustrators: Lisa Starr, Gary Head,
ScEYEnce Studios

Text Designer: Lisa Starr

Cover Designer: Bethany Casey

Cover and Title Page Image:
© Pete Oxford/Minden Pictures

Butterflies sip the tears of a yellow-spotted river
turtle sunning itself in Yasuní National Park,
Ecuador. Turtle tears supply the butterflies with
sodium, an essential nutrient missing from their
flower nectar diet in the Amazon rainforest.
Butterflies are almost never observed sipping
turtle tears outside of this small region, which
is famous for having one of the most diverse
assortments of species in the world. Currently,
oil drilling operations are destroying the forest
and wildlife in the park.

Compositor: Lachina Publishing Services

For product information and technology assistance, contact us at
Cengage Learning Customer & Sales Support, 1-800-354-9706.

For permission to use material from this text or product,
submit all requests online at **www.cengage.com/permissions**.
Further permissions questions can be emailed to
permissionrequest@cengage.com.

Library of Congress Control Number: 2014944587

ISBN-13: 978-1-305-25125-0

ISBN-10: 1-305-25125-3

Cengage Learning
20 Channel Center Street
Boston, MA 02210
USA

Cengage Learning is a leading provider of customized learning solutions with office
locations around the globe, including Singapore, the United Kingdom, Australia, Mexico,
Brazil, and Japan. Locate your local office at:
www.cengage.com/global.

Cengage Learning products are represented in Canada by Nelson Education, Ltd.

To learn more about Cengage Learning Solutions, visit **www.cengage.com.**

Purchase any of our products at your local college store or at our preferred
online store **www.cengagebrain.com**.

Printed in Canada
Print Number: 01 Print Year: 2014

Contents in Brief

Highlighted chapters are not included in *Evolution of Life*.

Detailed Contents

Preface

This edition of *Biology: The Unity and Diversity of Life* includes a wealth of new information reflecting recent discoveries in biology (details can be found in the *Power Bibliography*, which lists journal articles and other references used in the revision process; available upon request). Descriptions of current research, along with photos and videos of scientists who carry it out, underscore the concept that science is an ongoing endeavor carried out by a diverse community of people. Discussions include not only what was discovered, but also how the discoveries were made, how our understanding has changed over time, and what remains to be discovered. These discussions are provided in the context of a thorough, accessible introduction to well-established concepts and principles that underpin modern biology. Every topic is examined from an evolutionary perspective, emphasizing the connections between all forms of life.

Throughout the book, text and art have been revised to help students grasp difficult concepts. This edition also continues to focus on real world applications pertaining to the field of biology, including social issues arising from new research and developments. This edition covers in detail the many ways in which human activities are continuing to alter the environment and threaten both human health and Earth's biodiversity.

Changes to this Edition

Here are a few highlights of the revisions to this edition.

1 Invitation to Biology Renewed and updated emphasis on the relevance of new species discovery and the process of science.

2 Life's Chemical Basis New graphics illustrate elements and radioactive decay.

3 Molecules of Life New figure illustrates protein domains.

4 Cell Structure and Function New table summarizing cell theory; new photos of prokaryotes. Comparison of microscopy techniques updated using *Paramecium*. New figure shows food vacuoles in *Nassula*.

5 Ground Rules of Metabolism Temperature-dependent enzyme activity now illustrated with polymerases. New art and photos illustrate coenzymes, adhesion proteins, membrane trafficking, and energy transfer in redox reactions.

6 Where It Starts—Photosynthesis New photos illustrate phycobilins, stomata, adaptations of C4 plants, ice core sampling, smog in China. Light-dependent reactions art simplified.

7 How Cells Release Chemical Energy New photos illustrate mitochondrial disease and aerobic respiration.

8 DNA Structure and Function Concepts and illustrations of DNA hybridization and primers added to replication section. New photo of mutations caused by radiation at Chernobyl; new illustration of mutation.

9 From DNA to Protein Expanded material on the effects of mutation includes discussion of hairlessness in cats and a new micrograph of a sickled blood cell.

10 Gene Control New photos show transcription factors, X chromosome inactivation; new material explains evolution of lactose tolerance. New critical thinking question requires understanding of the effects of floral identity gene mutations.

11 How Cells Reproduce New photos illustrate mitosis, the mitotic spindle, and telomeres.

12 Meiosis and Sexual Reproduction New material on asexuality in mud snails and bdelloid rotifers. New micrograph shows multiple crossovers.

13 Observing Patterns in Inherited Traits New material about environmental effects on hemoglobin gene expression in *Daphnia*. New photos illustrate continuous variation.

14 Chromosomes and Human Inheritance Material on Tay-Sachs has been moved to this chapter as an illustration of autosomal recessive inheritance.

15 Studying and Manipulating Genomes Coverage of personal genetic testing updated with new medical applications, including the social impact of Angelina Jolie's response to her test. New photos of genetically modified animals. New "who's the daddy" critical thinking question offers students an opportunity to analyze a paternity test based on SNPs.

16 Evidence of Evolution New MRI showing coccyx illustrates a vestigial structure. Photos of 19th century naturalists added to emphasize the process of science that led to natural selection theory. Expanded coverage of fossil formation includes how banded iron formations provide evidence of the evolution of photosynthesis.

17 Processes of Evolution New opening essay on resistance to antibiotics as an outcome of agricultural overuse (warfarin material moved to illustrate directional selection). New art illustrates founder effect, and hypothetical example in text replaced with reduced diversity of *ABO* alleles in Native Americans. New art illustrates stasis in coelacanths.

18 Organizing Information About Species New material on DNA barcoding added to biochemical comparisons section. Data analysis activity revised to incorporate new data on honeycreeper ancestry.

19 Life's Origin and Early Evolution Added material about new discovery of 3.4-billion-year old fossil bacteria. New graphic illustrates endosymbiotic origin of mitochondria and chloroplasts.

20 Viruses, Bacteria, and Archaea Added information about Ebola and West Nile viruses, and newly discovered giant viruses.

21 Protists—The Simplest Eukaryotes New graphic depicts primary and secondary endosymbiosis. Added information about diatoms as a source of oil.

22 The Land Plants New essay about seed banks and the importance of sustain plant biodiversity.

23 Fungi More extensive coverage of fungal ecology; added information about white nose syndrome, a fungal disease of bats.

24 Animal Evolution—Invertebrates Updated information of medicines from invertebrates. New photos of terrestrial flatworm, plant-infecting roundworm.

25 Animal Evolution—Vertebrates Improved discussion of transition to land, with new illustration. Reorganized coverage of mammal evolution and diversity.

26 Human Evolution Updated to include latest discoveries about *Australopithecus sediba*, Denisovans, and Neanderthals.

27 Plant Tissues Carbon sequestration essay revised to include new data on wood production by old-growth redwoods. Reorganized to consolidate primary growth into its own section. Many new photos illustrate stem, leaf, and root structure. Material on fire scars added to section on dendroclimatology.

28 Plant Nutrition and Transport Illustration of Casparian strip integrated with new micrograph. Revisited section discusses phytoremediation at Ford's Rouge Center.

29 Life Cycles of Flowering Plants Updated material reflects current research on bee pollination behavior and colony collapse. New photos illustrate pollinators, fruit classification, asexual reproduction.

30 Communication Strategies in Plants Updates reflect ongoing major breakthroughs in the field of plant hormone function. New photos show apical dominance, effect of gibberellin, and abscission.

31 Animal Tissues and Organ Systems Added information about tissue regeneration in nonhuman animals; updated information about use of human and embryonic stem cells. Added information about blubber as a specialized adipose tissue.

32 Neural Control New opening essay about the effects of concussion on the brain. Reorganized coverage of psychoactive drugs. Added information about epidural anesthesia. Updated, improved coverage of memory.

33 Sensory Perception New opening essay about cochlear implants; revisited section discusses retinal implants, artificial limbs. Updated information about human sense of taste.

34 Endocrine Control Updated discussion of endocrine disruptors. New examples of pituitary gigantisms, dwarfism. Added information about role of melatonin in seasonal coat color changes.

35 Structural Support and Movement Added information about myostatin polymorphism in race horses to opening essay. New section discusses principles of animal location. Added information about boneless muscular organs such as the tongue.

36 Circulation More extensive coverage of plasma components. Discussion of genetics of blood types deleted. Improved coverage of and illustration of capillary exchange. Added information about blood pressure and jugular vein valves in giraffes.

37 Immunity Updated material on HIV/AIDS treatment strategies. New photos show T cell/APC interaction, skin as a surface barrier, a cytotoxic T cell killing a cancer cell, contact allergy, and victims of HIV.

38 Respiration Improved comparison of water and air as respiratory media with accompanying figure. Revised figure depicting first aid for choking victims to reflect latest guidelines. Discussion of human adaptation to high altitude now compares mechanisms in Tibetan and Andean populations.

39 Digestion and Nutrition New graphic depicting functional variations in animal dentition. New figure showing arrangement of organs that empty into the small intestine. Improved discussion of vitamin and mineral functions. New MRI illustrates how abdominal fat compresses internal organs. Added information about basal metabolic rate.

40 Maintaining the Internal Environment New subsection about climate-related adaptations in human populations.

41 Animal Reproductive Systems Coverage of intersex conditions dropped. Opening essay now discusses reproductive technology (IVF, egg banking); Revisited section discusses sperm banks. New section discusses location of animal gonads and the general mechanism of gamete formation. Reproductive function of human females now discussed before that of males; improved figure depicting the ovarian cycle.

42 Animal Development New opening essay about human birth defects, with a focus on cleft lip and palate. Improved photos illustrating apoptosis in digit development. Reorganized coverage of early human development. Added information about surgical delivery (cesarean section).

43 Animal Behavior Opening essay about effects of noise pollution on animal communication moved here and updated to reflect recent research. Revised discussion of the possible benefits of grouping.

44 Population Ecology Improved presentation of effects of predation on guppy life history. Revised, updated graphics.

45 Community Ecology Added information about and a photo of a brood parasite of ants. Added photo of the keystone species Pisaster.

46 Ecosystems More extensive discussion of aquifer depletion, salination; added information about ecological effects of over-allocation of river water. Updated discussion of the rise in atmospheric CO_2.

47 The Biosphere New opening essay about how winds and ocean currents distributed and are distributing material from the 2011 earthquake and tidal wave that affected Japan. Discussion of El Nino now a subsection within the chapter.

48 Human Impacts on the Biosphere New graphics of extinct animals: mastadon and dodo. Added information about and photo of endangered Florida lichen; added information about the Great Pacific Garbage Patch. Updated coverage of ozone depletion and effects of global climate change.

Student and Instructor Resources

Cengage Learning Testing Powered by Cognero
is a flexible, online system that allows you to:

• author, edit, and manage test bank content from multiple Cengage Learning solutions
• create multiple test versions in an instant
• deliver tests from your LMS, your classroom or wherever you want

Instructor Companion Site Everything you need for your course in one place! This collection of book-specific lecture and class tools is available online via www.cengage.com/login. Access and download PowerPoint presentations, images, instructor's manual, videos, and more

Cooperative Learning Cooperative Learning: Making Connections in General Biology, 2nd Edition, authored by Mimi Bres and Arnold Weisshaar, is a collection of separate, ready-to-use, short cooperative activities that have broad application for first year biology courses. They fit perfectly with any style of instruction, whether in large lecture halls or flipped classrooms. The activities are designed to address a range of learning objectives such as reinforcing basic concepts, making connections between various chapters and topics, data analysis and graphing, developing problem solving skills, and mastering terminology. Since each activity is designed to stand alone, this collection can be used in a variety of courses and with any text.

MindTap A personalized, fully online digital learning platform of authoritative content, assignments, and services that engages students with interactivity while also offering instructors their choice in the configuration of coursework and enhancement of the curriculum via web-apps known as MindApps. MindApps range from ReadSpeaker (which reads the text out loud to students), to Kaltura (allowing you to insert inline video and audio into your curriculum). MindTap is well beyond an eBook, a homework solution or digital supplement, a resource center website, a course delivery platform, or a Learning Management System. It is the first in a new category —the Personal Learning Experience.

New for this edition! MindTap has an integrated Study Guide, expanded quizzing and application activities, and an integrated Test Bank.

Aplia for Biology The Aplia system helps students learn key concepts via Aplia's focused assignments and active learning opportunities that include randomized, automatically graded questions, exceptional text/art integration, and immediate feedback. Aplia has a full course management system that can be used independently or in conjunction with other course management systems such as MindTap, D2L, or Blackboard.

Acknowledgments

Writing, revising, and illustrating a biology textbook is a major undertaking for two full-time authors, but our efforts constitute only a small part of what is required to produce and distribute this one. We are truly fortunate to be part of a huge team of very talented people who are as committed as we are to creating and disseminating an exceptional science education product.

Biology is not dogma; paradigm shifts are a common outcome of the fantastic amount of research in the field. Ideas about what material should be taught and how best to present that material to students changes even from one year to the next. It is only with the ongoing input of our many academic reviewers and advisors (see opposite page) that we can continue to tailor this book to the needs of instructors and students while integrating new information and models. We continue to learn from and be inspired by these dedicated educators. A special thanks goes to Jose Panero for his extensive and detailed review for this edition.

On the production side of our team, the indispensable Grace Davidson orchestrated a continuous flow of files, photos, and illustrations while managing schedules, budgets, and whatever else happened to be on fire at the time. Grace, thank you as always for your patience and dedication. Thank you also to Cheryl DuBois, John Sarantakis, and Christine Myaskovsky for your help with photoresearch. Copyeditor Anita Hueftle and proofreader Kathy Dragolich, your valuable suggestions kept our text clear and concise.

Yolanda Cossio, thank you for continuing to support us and for encouraging our efforts to innovate and improve. Peggy Williams, we are as always grateful for your enthusiastic, thoughtful guidance, and for your many travels (and travails) on behalf of our books.

Thanks to Hal Humphrey our Cengage Production Manager, Tom Ziolkowski our Marketing Manager, Lauren Oliveira who creates our exciting technology package, Associate Content Developers Casey Lozier and Kellie Petruzzelli, and Product Assistant Victor Luu.

Lisa Starr and Christine Evers, May 2014

Influential Class Testers and Reviewers

Brenda Alston-Mills
North Carolina State University

Kevin Anderson
Arkansas State University - Beebe

Norris Armstrong
University of Georgia

Tasneem Ashraf
Coshise College

Dave Bachoon
Georgia College & State University

Neil R. Baker
The Ohio State University

Andrew Baldwin
Mesa Community College

David Bass
University of Central Oklahoma

Lisa Lynn Boggs
Southwestern Oklahoma State
University

Gail Breen
University of Texas at Dallas

Marguerite "Peggy" Brickman
University of Georgia

David Brooks
East Central College

David William Bryan
Cincinnati State College

Lisa Bryant
Arkansas State University - Beebe

Katherine Buhrer
Tidewater Community College

Uriel Buitrago-Suarez
Harper College

Sharon King Bullock
Virginia Commonwealth University

John Capehart
University of Houston - Downtown

Daniel Ceccoli
American InterContinental
University

Tom Clark
Indiana University South Bend

Heather Collins
Greenville Technical College

Deborah Dardis
Southeastern Louisiana University

Cynthia Lynn Dassler
The Ohio State University

Carole Davis
Kellogg Community College

Lewis E. Deaton
University of Louisiana - Lafayette

Jean Swaim DeSaix
University of North Carolina -
Chapel Hill

(Joan) Lee Edwards
Greenville Technical College

Hamid M. Elhag
Clayton State University

Patrick Enderle
East Carolina University

Daniel J. Fairbanks
Brigham Young University

Amy Fenster
Virginia Western Community
College

Kathy E. Ferrell
Greenville Technical College

Rosa Gambier
Suffok Community College -
Ammerman

Tim D. Gaskin
Cuyahoga Community College -
Metropolitan

Stephen J. Gould
Johns Hopkins University

Laine Gurley
Harper College

Marcella Hackney
Baton Rouge Community College

Gale R. Haigh
McNeese State University

John Hamilton
Gainesville State

Richard Hanke
Rose State Community College

Chris Haynes
Shelton St. Community College

Kendra M. Hill
South Dakota State University

Juliana Guillory Hinton
McNeese State University

W. Wyatt Hoback
University of Nebraska, Kearney

Kelly Hogan
University of North Carolina

Norma Hollebeke
Sinclair Community College

Robert Hunter
Trident Technical College

John Ireland
Jackson Community College

Thomas M. Justice
McLennan College

Timothy Owen Koneval
Laredo Community College

Sherry Krayesky
University of Louisiana -
Lafayette

Dubear Kroening
University of Wisconsin - Fox
Valley

Jerome Krueger
South Dakota State University

Jim Krupa
University of Kentucky

Mary Lynn LaMantia
Golden West College

Dale Lambert
Tarrant County College

Kevin T. Lampe
Bucks County Community College

Susanne W. Lindgren
Sacramento State University

Madeline Love
New River Community College

Dr. Kevin C. McGarry
Kaiser College - Melbourne

Ashley McGee
Alamo College

Jeanne Mitchell
Truman State University

Alice J. Monroe
St. Petersburg College -
Clearwater

Brenda Moore
Truman State University

Erin L. G. Morrey
Georgia Perimeter College

Rajkumar "Raj" Nathaniel
Nicholls State University

Francine Natalie Norflus
Clayton State University

Harold Olivey
Indiana University Northwest

Alexander E. Olvido
Virginia State University

John C. Osterman
University of Nebraska, Lincoln

Jose L. Panero
University of Texas

Bob Patterson
North Carolina State University

Shelley Penrod
North Harris College

Carla Perry
Community College of Philadelphia

Mary A. (Molly) Perry
Kaiser College - Corporate

John S. Peters
College of Charleston

Carlie Phipps
SUNY IT

Michael Plotkin
Mt. San Jacinto College

Ron Porter
Penn State University

Karen Raines
Colorado State University

Larry A. Reichard
Metropolitan Community College -
Maplewood

Jill D. Reid
Virginia Commonwealth University

Robert Reinswold
University of Northern Colorado

Ashley E. Rhodes
Kansas State University

David Rintoul
Kansas State University

Darryl Ritter
Northwest Florida State College

Amy Wolf Rollins
Clayton State University

Sydha Salihu
West Virginia University

Jon W. Sandridge
University of Nebraska

Robin Searles-Adenegan
Morgan State University

Erica Sharar
IVC; National University

Julie Shepker
Kaiser College - Melbourne

Rainy Shorey
Illinois Central College

Eric Sikorski
University of South Florida

Phoebe Smith
Suffolk County Community College

Robert (Bob) Speed
Wallace Junior College

Tony Stancampiano
Oklahoma City Community College

Jon R. Stoltzfus
Michigan State University

Peter Svensson
West Valley College

Jeffrey L. Travis
University at Albany

Nels H. Troelstrup, Jr.
South Dakota State University

Allen Adair Tubbs
Troy University

Will Unsell
University of Central Oklahoma

Rani Vajravelu
University of Central Florida

Jack Waber
West Chester University of
Pennsylvania

Kathy Webb
Bucks County Community College

Amy Stinnett White
Virginia Western Community
College

Virginia White
Riverside Community College

Robert S. Whyte
California University of
Pennsylvania

Kathleen Lucy Wilsenn
University of Northern Colorado

Penni Jo Wilson
Cleveland State Community
College

Robert Wise
University of Wisconsin Oshkosh

Michael L. Womack
Macon State College

Maury Wrightson
Germanna Community College

Mark L. Wygoda
McNeese State University

Lan Xu
South Dakota State University

Poksyn ("Grace") Yoon
Johnson and Wales University

Muriel Zimmermann
Chaffey College

LEARNING ROADMAP

You may wish to review critical thinking (Section 1.6) before reading this chapter, which explores a clash between belief and science (1.9). We revisit radioisotopes (2.2), the effect of photosynthesis on Earth's early atmosphere (7.1), the genetic code and mutations (9.4, 9.6), master genes (10.3, 10.4), alleles (12.2), and evolution by gene duplication (14.5).

EMERGENCE OF EVOLUTIONARY THOUGHT

Nineteenth-century naturalists investigating the global distribution of species discovered patterns that could not be explained within the framework of traditional belief systems.

A THEORY TAKES FORM

Evidence of evolution, or change in lines of descent, led Charles Darwin and Alfred Wallace to develop a theory of how traits that define each species change over time.

EVIDENCE FROM FOSSILS

The fossil record provides physical evidence of past changes in many lines of descent. We use the property of radioisotope decay to determine the age of rocks and fossils.

INFLUENTIAL GEOLOGIC FORCES

Over millions of years, slow movements of Earth's outer crust have affected land and oceans. The changes have profoundly influenced life's evolution.

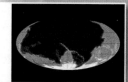

THE GEOLOGIC TIME SCALE

By studying rock layers and fossils in them, we can correlate geologic and evolutionary events. The correlation helps explain the distribution of species, past and present.

The next two chapters continue the theme of evolutionary processes, including how natural selection works (17.4–17.7), how the movement of tectonic plates can affect evolution (Section 17.10), and what comparative morphology (18.3) and DNA sequence comparisons (18.4) can tell us about shared evolutionary history. A continent's climate is affected by its position on the globe, as you will see in Chapter 47.

How do you think about time? Perhaps you can conceive of a few hundred years of human events, maybe a few thousand, but how about a few million? Envisioning the very distant past requires an intellectual leap from the familiar to the unknown. One way to make that leap involves, surprisingly, asteroids. Asteroids are small planets hurtling through space. They range in size from 1 to 1,500 kilometers (roughly 0.5 to 1,000 miles) across. Millions of them orbit the sun between Mars and Jupiter—cold, stony leftovers from the formation of our solar system. Asteroids are difficult to see even with the best telescopes, because they do not emit light. Many cross Earth's orbit, but most of those pass us by before we know about them. Some have not passed us at all.

Consider the mile-wide Barringer Crater in Arizona (**FIGURE 16.1A**). An asteroid 45 meters (150 feet) wide made this impressive pockmark in the desert sandstone when it slammed into Earth 50,000 years ago with an impact 150 times more powerful than the bomb that leveled Hiroshima. No humans were in North America at the time of the impact. If there were no witnesses, how is it possible to know anything about what happened? We often reconstruct history by studying physical evidence of past events. Geologists were able to infer the most probable cause of the Barringer Crater by analyzing tons of meteorites, melted sand, and other rocky clues at the site.

Similar evidence points to even larger asteroid impacts in the more distant past. For example, fossil hunters have long known about a mass extinction, or permanent loss of major groups of organisms, that occurred 66 million years ago. The event is marked by an unusual, worldwide formation of sedimentary rock (**FIGURE 16.1B**). This formation is called the K–Pg boundary sequence (it was formerly known as the K–T boundary). There are plenty of dinosaur fossils below this formation. Above it, in rock layers that were deposited more recently, there are no dinosaur fossils, anywhere. A gigantic impact crater—274 kilometers (170 miles) across and 1 kilometer (3,000 feet) deep—off the coast of what is now the Yucatán Peninsula dates to about 66 million years ago. Coincidence? Many scientists say no. The asteroid that made the Yucatán crater had to be at least 20 kilometers (12 miles) wide when it hit. The impact of an asteroid that size would have been *40 million times* more powerful than the one that made the Barringer Crater. The scientists infer that the impact caused a global catastrophe of sufficient scale to wipe out the dinosaurs.

You are about to make an intellectual leap through time to consider events that were not known even a few centuries ago. We invite you to launch yourself from this premise: Natural phenomena that occurred in the past can be explained by the very same physical, chemical, and biological processes that operate today. That premise is the foundation for scientific research into the history of life. The research represents a shift from experience to inference—from the known to what can only be surmised—and it gives us astonishing glimpses into the distant past.

A What made the Barringer Crater in Arizona? Rocky evidence points to a 300,000-ton asteroid that collided with Earth 50,000 years ago.

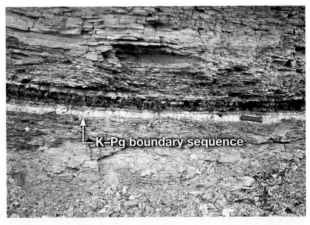

B The K–Pg boundary sequence, an unusual, worldwide formation of sedimentary rock that formed 66 million years ago. This formation marks an abrupt transition in the fossil record that implies a mass extinction. The red pocketknife gives an idea of scale.

FIGURE 16.1 Evidence to inference.

✔ Belief systems are influenced by the extent of our understanding of the natural world. Those that are inconsistent with systematic observations tend to change over time.

About 2,300 years ago, the Greek philosopher Aristotle described nature as a continuum of organization, from lifeless matter through complex plants and animals. Aristotle's work greatly influenced later European thinkers, who adopted his view of nature and modified it in light of their own beliefs. By the fourteenth century, Europeans generally believed that a "great chain of being" extended from the lowest form (snakes), up through humans, to spiritual beings. Each link in the chain was a species, and each was said to have been forged at the same time, in one place, and in a perfect state. The chain itself was complete and continuous. Because everything that needed to exist already did, there was no room for change.

In the 1800s, European naturalists embarked on globe-spanning survey expeditions and brought back tens of thousands of plants and animals from Asia, Africa, North and South America, and the Pacific Islands. Each newly discovered species was carefully catalogued as another link in the chain of being. The explorers began to see patterns in where species live and similarities in body plans, and had started to think about the natural forces that shape life. These explorers were pioneers in **biogeography**, the study of patterns in the geographic distribution of species and communities. Some of the patterns raised questions that could not be answered within the framework of prevailing belief systems. For example, globe-trotting explorers had discovered plants and animals living in extremely isolated places. The isolated species looked suspiciously similar to species living on the other side of impassable mountain ranges, or across vast expanses

FIGURE 16.3 Similar-looking, unrelated species. On the left, an African milk barrel cactus (*Euphorbia horrida*), native to the Great Karoo desert of South Africa. On the right, saguaro cactus (*Carnegiea gigantea*), native to the Sonoran Desert of Arizona.

of open ocean. Consider the emu, rhea, and ostrich, three types of bird native to three different continents (**FIGURE 16.2**). These birds share a set of unusual features. All are very large, with long, muscular legs and necks. All are also flightless, sprinting about in flat, open grasslands about the same distance from the equator. Alfred Wallace, an explorer particularly interested in the geographical distribution of animals, thought that the shared traits might mean that the birds descended from a common ancestor (and he was correct), but he had no idea how they could have ended up on different continents.

Naturalists of the time also had trouble classifying organisms that are very similar in some features, but different in others. For example, both plants shown in **FIGURE 16.3** live in hot deserts where water is seasonally scarce. Both have rows of sharp spines that deter herbivores, and both store water in their thick, fleshy

A Emu, native to Australia   **B** Rhea, native to South America   **C** Ostrich, native to Africa

FIGURE 16.2 Similar-looking, related animals native to distant geographic realms. These birds are unlike most others in several unusual features, including long, muscular legs and an inability to fly. All are native to open grassland regions about the same distance from the equator.

stems. However, their reproductive parts are very different, so these plants cannot be (and are not) as closely related as their outward appearance might suggest.

Observations such as these are examples of **comparative morphology**, the study of anatomical patterns: similarities and differences among the body plans of organisms. Today, comparative morphology is only one branch of taxonomy (Section 1.5), but in the nineteenth century it was the only way to distinguish species. In some cases, comparative morphology revealed anatomical details (body parts with no apparent function, for example) that added to the mounting confusion. If every species had been created in a perfect state, then why were there useless parts such as wings in birds that do not fly, eyes in moles that are blind, or remnants of a tail in humans (**FIGURE 16.4**)?

Fossils were puzzling too. A **fossil** is physical evidence—remains or traces—of an organism that lived in the ancient past. Geologists mapping rock formations exposed by erosion or quarrying had discovered identical sequences of rock layers in different parts of the world. Deeper layers held fossils of simple marine life. Layers above those held similar but more complex fossils (**FIGURE 16.5**). In higher layers, fossils that were similar but even more complex resembled modern species. What did these sequences mean? Fossils of many animals unlike any living ones were also being unearthed. If these animals had been perfect at the time of creation, then why had they become extinct?

Taken as a whole, the accumulating discoveries from biogeography, comparative morphology, and geology did not fit with prevailing beliefs of the nineteenth century. If species had not been created in a perfect state (and extinct species, fossil sequences, and "useless" body parts implied that they had not), then perhaps species had indeed changed over time.

---

biogeography Study of patterns in the geographic distribution of species and communities.
comparative morphology The scientific study of similarities and differences in body plans.
fossil Physical evidence of an organism that lived in the ancient past.

TAKE-HOME MESSAGE 16.2
Why did observations of nature change our thinking in the nineteenth century?

✔ Increasingly extensive observations of nature in the nineteenth century did not fit with prevailing belief systems.

✔ Cumulative findings from biogeography, comparative morphology, and geology led naturalists to question traditional ways of interpreting the natural world.

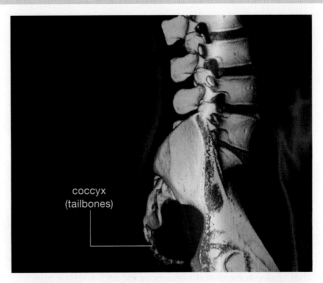

coccyx (tailbones)

FIGURE 16.4 A vestigial structure: human tailbones. Nineteenth-century naturalists were well aware of—but had trouble explaining—body structures such as human tailbones that had apparently lost most or all function.

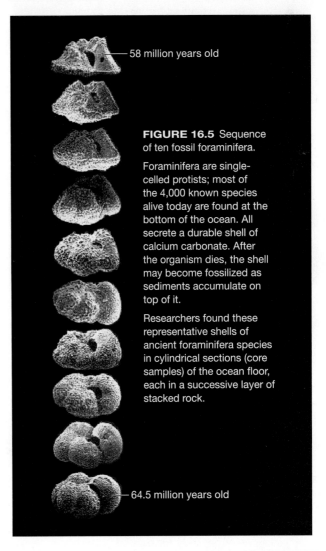

58 million years old

FIGURE 16.5 Sequence of ten fossil foraminifera.

Foraminifera are single-celled protists; most of the 4,000 known species alive today are found at the bottom of the ocean. All secrete a durable shell of calcium carbonate. After the organism dies, the shell may become fossilized as sediments accumulate on top of it.

Researchers found these representative shells of ancient foraminifera species in cylindrical sections (core samples) of the ocean floor, each in a successive layer of stacked rock.

64.5 million years old

✔ In the 1800s, many naturalists realized that life on Earth had changed over time, and began to think about what could have caused the changes.

Squeezing New Evidence Into Old Beliefs

In the nineteenth century, naturalists were faced with increasing evidence that life on Earth, and even Earth itself, had changed over time. Around 1800, Georges Cuvier (left), an expert in zoology and paleontology, was trying to make sense of the new information. He had observed abrupt changes in the fossil record, and knew that many fossil species seemed to have no living counterparts. Given this evidence, he proposed an idea startling for the time: Many species that had once existed were now extinct. Cuvier also knew about evidence that Earth's surface had changed.

For example, he had seen fossilized seashells in rocks at the tops of mountains far from modern seas. Like most others of his time, he assumed Earth's age to be in the thousands, not billions, of years. He reasoned that geologic forces unlike any known at the time would have been necessary to raise seafloors to mountaintops in this short time span. Catastrophic geological events would have caused extinctions, after which surviving species repopulated the planet. Cuvier's idea came to be known as **catastrophism**. We now know it is incorrect; geologic processes have not changed over time.

Another naturalist, Jean-Baptiste Lamarck, was thinking about processes that might drive **evolution**, or change in a line of descent. A line of descent is also

called a **lineage**. Lamarck (left) thought that a species gradually improved over generations because of an inherent drive toward perfection, up the chain of being. The drive directed an unknown "fluida" into body parts needing change. By Lamarck's hypothesis, environmental pressures cause an internal requirement for change in an individual's body, and the resulting change is inherited by offspring.

Try using Lamarck's hypothesis to explain why a giraffe's neck is very long. We might predict that some short-necked ancestor of the modern giraffe stretched its neck to browse on leaves beyond the reach of other animals. The stretches may have even made its neck a bit longer. By Lamarck's hypothesis, that animal's offspring would inherit a longer neck. The modern giraffe would have been the result of many generations that strained to reach ever loftier leaves. Lamarck was correct in thinking that environmental factors affect a species' traits, but his understanding of how inheritance works was incomplete.

Darwin and the HMS *Beagle*

Lamarck's ideas about evolution influenced the thinking of Charles Darwin, who, at the age of 22, joined a survey expedition to South America on the ship *Beagle*. Since he was eight years old, Darwin had wanted to hunt, fish, collect shells, or watch insects and birds—anything but sit in school. After a failed attempt to study medicine in college, he earned a degree in theol-

FIGURE 16.6 Voyage of the HMS *Beagle*. With Darwin aboard as ship's naturalist, the vessel (top) originally set sail to map the coast of South America, but ended up circumnavigating the globe (bottom). The path of the voyage is shown from red to blue. Darwin's detailed observations of the geology, fossils, plants, and animals he encountered on this expedition changed the way he thought about evolution.

catastrophism Now-abandoned hypothesis that catastrophic geologic forces unlike those of the present day shaped Earth's surface.
evolution Change in a line of descent.
lineage Line of descent.
theory of uniformity Idea that gradual repetitive processes occurring over long time spans shaped Earth's surface.

CREDITS: (6) top, © Gordon Chancellor; bottom, © Cengage Learning.

ogy from Cambridge. During his studies, Darwin had spent most of his time with faculty members and other students who embraced natural history.

The *Beagle* set sail for South America in December 1831 (**FIGURE 16.6**). The young man who had hated school and had no formal training in science quickly became an enthusiastic naturalist. During the *Beagle*'s five-year voyage, Darwin found many unusual fossils, and saw diverse species living in environments that ranged from the sandy shores of remote islands to plains high in the Andes. Along the way, he read the first volume of a new and popular book, Charles Lyell's *Principles of Geology*. Lyell (left) was a proponent of what became known as the **theory of uniformity**, the idea that gradual, repetitive change had shaped Earth. For many years, geologists had been chipping away at the sandstones, limestones, and other types of rocks that form from accumulated sediments at the bottom of lakes, rivers, and oceans. These rocks held evidence that gradual processes of geologic change operating in the present were the same ones that operated in the distant past.

By the theory of uniformity, strange catastrophes were not necessary to explain Earth's surface. Gradual, everyday geologic processes such as erosion by wind and water could have sculpted Earth's current landscape over great spans of time. This theory challenged the prevailing belief in Europe that Earth was 6,000 years old. According to traditional scholars, people had recorded everything that happened in those 6,000 years—and in all that time, no one had mentioned seeing a species evolve. However, by Lyell's calculations, it must have taken millions of years to sculpt Earth's surface. Darwin's exposure to Lyell's ideas gave him insights into the geologic history of the regions he would encounter on his journey. Was millions of years enough time for species to evolve? Darwin thought that it was (**FIGURE 16.7**).

TAKE-HOME MESSAGE 16.3

How did new evidence change the way people in the 19th century thought about the history of life?

✔ In the 1800s, fossils and other evidence led some naturalists to propose that Earth and the species on it had changed over time. The naturalists also began to reconsider the age of Earth.

✔ Darwin's detailed observations of nature during a five-year voyage around the world changed his ideas about how evolution occurs.

FIGURE 16.7 Charles Darwin and part of a page from his 1836 notes on the "Transmutation of Species."

The text reads as follows: "Let a pair be introduced and increase slowly, from many enemies, so as often to intermarry who will dare say what result. According to this view animals on separate islands ought to become different if kept long enough apart with slightly differing circumstances."

✔ Darwin's observations of species in different parts of the world helped him understand a driving force of evolution.

Among the thousands of specimens Darwin collected on his voyage and sent to England were fossil glypto-dons from Argentina. These armored mammals are extinct, but they have many traits in common with modern armadillos (**FIGURE 16.8**). For example, arma-dillos live only in places where glyptodons once lived. Like glyptodons, armadillos have helmets and protec-tive shells that consist of unusual bony plates. Could the shared traits mean that glyptodons were ancient relatives of armadillos? If so, perhaps traits of their common ancestor had changed in the line of descent that led to armadillos. But why would such changes have occurred?

A Key Insight—Variation in Traits

After Darwin returned to England, he pondered his notes and fossils, and read an essay by one of his con-temporaries, economist Thomas Malthus (left). Malthus had correlated increases in the size of human populations with epi-sodes of famine, disease, and war. He pro-posed the idea that humans run out of food, living space, and other resources because they tend to reproduce beyond the capacity of their environment to sustain them. When that happens, the individuals of a population must either compete with one another for the limited resources, or develop new technologies to increase productivity. Darwin realized that Malthus's ideas had wider application: All populations, not just human ones, must have the capacity to produce more individuals than their environment can support.

Reflecting on his journey, Darwin started thinking about how individuals of a species often vary a bit in the details of shared traits such as size, coloration, and so on. He saw such variation among finch spe-cies on isolated islands of the Galápagos archipelago. This island chain is separated from South America by 900 kilometers (550 miles) of open ocean, so most spe-cies living on the islands did not have the opportunity for interbreeding with mainland populations. The Galápagos island finches resembled finch species in South America, but many had unique traits that suited their particular island habitat.

Darwin was familiar with dramatic variations in traits that selective breeding could produce in pigeons, dogs, and horses. He recognized that a natural environ-ment could similarly select forms of traits that make individuals of a population suited to it. It dawned on

A Fossil of a glyptodon, an automobile-sized mammal that existed from 2 million to 15,000 years ago.

B A modern armadillo, about a foot long.

FIGURE 16.8 Ancient relatives: glyptodon and armadillo.

Even though these animals are widely separated in time, they share a restricted distribution and unusual traits, including a shell and hel-met of keratin-covered bony plates—a material similar to crocodile and lizard skin. (The fossil in **A** is missing its helmet.) Their unique shared traits were a clue that helped Darwin develop the theory of evolution by natural selection.

Darwin that having a particular form of a shared trait might give an individual an advantage over compet-ing members of its species. In any population, some individuals have forms of shared traits that make them better suited to their environment than others. In other words, individuals of a natural population vary in fitness. Today, we define **fitness** as the degree of adaptation to a specific environment, and measure it by relative genetic contribution to future generations. A form of a heritable trait that enhances an individual's fitness is called an evolutionary **adaptation**, or **adap-tive trait**. Over many generations, individuals that have adaptive traits tend to survive longer and reproduce more than their less fit rivals. Darwin understood that this process, which he called **natural selection**, could be a mechanism by which evolution occurs. If an indi-vidual has a form of a trait that makes it better suited to an environment, then it is better able to survive. If an individual is better able to survive, then it has

FIGURE 16.9 Alfred Wallace, codiscoverer of natural selection.

Table 16.1 Principles of Natural Selection

Observations About Populations

✔ Natural populations have an inherent capacity to increase in size over time.

✔ As a population expands, resources that are used by its individuals (such as food and living space) eventually become limited.

✔ When resources are limited, individuals of a population compete for them.

Observations About Genetics

✔ Individuals of a species share certain traits.

✔ Individuals of a natural population vary in the details of those shared traits.

✔ Shared traits have a heritable basis, in genes. Slightly different versions of those genes (alleles) give rise to variation in shared traits.

Inferences

✔ A certain form of a shared trait may make its bearer better able to survive.

✔ Individuals of a population that are better able to survive tend to leave more offspring.

✔ Thus, an allele associated with an adaptive trait tends to become more common in a population over time.

a better chance of living long enough to produce off-spring. If individuals with an adaptive form of a trait produce more offspring than those that do not, then the frequency of that form will tend to increase in the population over successive generations. **TABLE 16.1** summarizes this reasoning in modern terms.

Great Minds Think Alike

Darwin wrote out his ideas about natural selection, but let ten years pass without publishing them. In the meantime, Alfred Wallace (**FIGURE 16.9**), who had been studying wildlife in the Amazon basin and the Malay Archipelago, wrote an essay and sent it to Darwin for advice. Wallace's essay outlined evolution by natural selection—the very same theory as Darwin's. Wallace had written earlier letters to Darwin and Lyell about patterns in the geographic distribution of species, and had come to the same conclusion.

In 1858, just weeks after Darwin received Wallace's essay, the theory of evolution by natural selection was presented at a scientific meeting. Both Darwin and Wallace were credited as authors. Wallace was still in the field and knew nothing about the meeting, which Darwin did not attend. The next year, Darwin published *On the Origin of Species*, which laid out detailed evidence in support of the theory. Many people had already accepted the idea of descent with modification (evolution). However, there was a fierce debate over the idea that natural selection drives evolution. Decades would pass before experimental evidence from the field of genetics led to its widespread acceptance as a theory by the scientific community.

As you will see in the remainder of this unit, the theory of evolution by natural selection is supported by and helps explain the fossil record as well as similarities and differences in the form, function, and biochemistry of living things.

---

adaptation (**adaptive trait**) A form of a heritable trait that enhances an individual's fitness in a particular environment.
fitness Degree of adaptation to an environment, as measured by an individual's relative genetic contribution to future generations.
natural selection Differential survival and reproduction of individuals of a population based on differences in shared, heritable traits.

TAKE-HOME MESSAGE 16.4

What is natural selection?

✔ Natural selection is a process that drives evolutionary change: Individuals of a population survive and reproduce with differing success depending on the details of their shared, heritable traits.

✔ Traits favored in a particular environment are adaptive.

CREDITS: (9) Down House and The Royal College of Surgeons of England; (Table 16.1) © Cengage Learning.

✔ The fossil record holds clues to life's evolution, but it will always be incomplete.

A Fossil skeleton of an ichthyosaur that lived about 200 million years ago. These marine reptiles were about the same size as modern porpoises, breathed air like them, and probably swam as fast, but the two groups are not closely related.

B Extinct wasp encased in amber, which is ancient tree sap. This 9-mm-long insect lived about 20 million years ago.

C Fossilized leaf from a 260-million-year-old *Glossopteris*, a type of plant called a seed fern.

D Fossilized footprints of a theropod, a name that means "beast foot." This group of carnivorous dinosaurs, which includes the familiar *Tyrannosaurus rex*, arose about 250 million years ago.

E Coprolite (fossilized feces). Fossilized food remains and parasitic worms inside coprolites offer clues about the diet and health of extinct species. A foxlike animal excreted this one.

FIGURE 16.10 Examples of fossils.

Even before Darwin's time, fossils were recognized as stone-hard evidence of earlier forms of life. Most fossils consist of mineralized bones, teeth, shells, seeds, spores, or other body parts (**FIGURE 16.10A–C**). Trace fossils such as footprints and other impressions, nests, burrows, trails, eggshells, or feces are evidence of an organism's activities (**FIGURE 16.10D,E**).

The process of fossilization typically begins when an organism or its traces become covered by sediments, mud, or ash. Groundwater then seeps into the remains, filling spaces around and inside of them. Minerals dissolved in the water gradually replace minerals in bones and other hard tissues. Mineral particles that crystallize and settle out of the groundwater inside cavities and impressions form detailed imprints of internal and external structures. Sediments that slowly accumulate on top of the site exert increasing pressure, and, after a very long time, extreme pressure transforms the mineralized remains into rock.

Most fossils are found in layers of sedimentary rock (**FIGURE 16.11**). Sedimentary rocks form as rivers wash silt, sand, volcanic ash, and other materials from land to sea. Mineral particles in the materials settle on seafloors in horizontal layers that vary in thickness and composition. After hundreds of millions of years, the layers of sediments become compacted into layers of rock. Even though most sedimentary rock forms at the bottom of a sea, geologic processes can tilt the rock and lift it far above sea level, where the layers may become exposed by the erosive forces of water and wind.

Biologists study sedimentary rock formations in order to understand life's historical context. Features of the formations can provide information about conditions  in the environment in which they formed. Consider banded iron, a unique formation named after its distinctive striped appearance (left). Huge deposits of this sedimentary rock are the source of most iron we mine for steel today, but they also hold a record of how the evolution of the noncyclic pathway of photosynthesis changed the chemistry of Earth. Banded iron started forming about 2.4 billion years ago, right after photosynthesis evolved (Section 7.1). At that time, Earth's atmosphere and ocean contained very little oxygen, so almost all of the iron on Earth was in a reduced form (Section 5.5). Reduced iron dissolves in water, and ocean water contained a lot of it. Oxygen released into the ocean by early photosynthetic bacteria quickly combined with the dissolved iron. The resulting oxidized iron compounds are completely insoluble in water, and

CREDITS: (10A) Jonathan Blair; (10B) © Dr. Michael Engel, University of Kansas; (10C) Martin Land/Science Source; (10D) © Pixtal/SuperStock; (10E) Courtesy of Stan Celestian/Glendale Community College Earth Science Image Archive; (in text) Natural History Museum, London/Science Photo Library/Science Source.

they began to rain down on the ocean floor in massive quantities. These compounds accumulated in sediments that would eventually become compacted into banded iron formations. This process continued for about 600 million years. After that, ocean water no longer contained very much dissolved iron, and oxygen gas bubbling out of it had oxidized the iron in rocks exposed to the atmosphere.

The Fossil Record

We have fossils for more than 250,000 known species. Considering the current range of biodiversity, there must have been many millions more, but we will never know all of them. Why not? The odds are against finding evidence of an extinct species, because fossils are relatively rare. Typically, when an organism dies, its remains are obliterated quickly by scavengers. Organic materials decompose in the presence of moisture and oxygen, so remains that escape scavenging endure only if they dry out, freeze, or become encased in an air-excluding material such as sap, tar, or mud. Remains that do become fossilized are usually crushed or scattered by erosion and other geologic assaults.

In order for us to know about an extinct species that existed long ago, we have to find a fossil of it. At least one specimen had to be buried before it decomposed or something ate it. The burial site had to escape destructive geologic events, and end up in a place that we can find today. Most ancient species had no hard parts to fossilize, so we do not find much evidence of them. For example, there are many fossils of bony fishes and mollusks with hard shells, but few fossils of the jellyfishes and soft worms that were probably much more common. Also think about relative numbers of organisms. Fungal spores and pollen grains are typically released by the billions. By contrast, the earliest humans lived in small bands and few of their offspring survived. The odds of finding even one fossilized human bone are much smaller than the odds of finding a fossilized fungal spore. Finally, imagine two species, one that existed only briefly and the other for billions of years. Which is more likely to be represented in the fossil record?

TAKE-HOME MESSAGE 16.5

What are fossils?

✔ Fossils are evidence of organisms that lived in the remote past, a stone-hard historical record of life.

✔ The fossil record will never be complete because fossils are relatively rare. It is slanted toward hard-bodied species that lived in large populations and persisted for a long time.

A Two types of sedimentary rock. Sandstones (left) consist of compacted grains of sand or minerals; shales (right) consist of ancient compacted clay or mud.

B Cindy Looy and Mark Sephton climb the walls of Butterloch Canyon, Italy, to look for fossilized fungal spores in a 251-million-year-old layer of sedimentary rock.

C A fossilized trilobite (an ancient marine relative of centipedes) found in a shale formation by fossil hunters in Yoho National Park, British Columbia.

FIGURE 16.11 Fossils are most often found in layered sedimentary rock. This type of rock forms over hundreds of millions of years, often at the bottom of a sea. Geologic processes can tilt the rock and lift it far above sea level, where the layers become exposed by the erosive forces of water and wind.

16.6 Filling In Pieces of the Puzzle

✔ Radiometric dating reveals the age of rocks and fossils.

✔ New fossil discoveries are continually filling gaps in our understanding of the ancient history of many lineages.

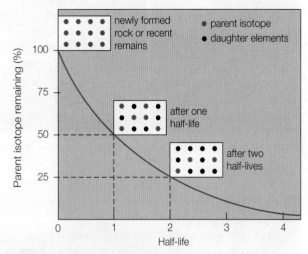

FIGURE 16.12 ▶**Animated** Half-life. **FIGURE IT OUT** How much of any radioisotope remains after two half-lives have passed?

Answer: 25 percent

A Long ago, ^{14}C and ^{12}C were incorporated into the tissues of a nautilus. The carbon atoms were part of organic molecules in the animal's food. ^{12}C is stable and ^{14}C decays, but the proportion of the two isotopes in the nautilus's tissues remained the same. Why? The nautilus continued to gain both types of carbon atoms in the same proportions from its food.

B The nautilus stopped eating when it died, so its body stopped gaining carbon. The ^{12}C atoms in its tissues were stable, but the ^{14}C atoms (represented as red dots) were decaying into nitrogen atoms. Thus, over time, the amount of ^{14}C decreased relative to the amount of ^{12}C. After 5,730 years, half of the ^{14}C had decayed; after another 5,730 years, half of what was left had decayed, and so on.

C Fossil hunters discover the fossil and measure its content of ^{14}C and ^{12}C. They use the ratio of these isotopes to calculate how many half-lives have passed since the organism died. For example, if its ^{14}C to ^{12}C ratio is one-eighth of the ratio in living organisms, then three half-lives $(1/2)^3$ must have passed since it died. Three half-lives of ^{14}C is 17,190 years.

FIGURE 16.13 ▶**Animated** Example of how radiometric dating can be used to find the age of a carbon-containing fossil. Carbon 14 (^{14}C) is a radioisotope of carbon that decays into nitrogen. It forms in the atmosphere and combines with oxygen to become CO_2, which enters food chains by way of photosynthesis.

Radiometric Dating

Remember from Section 2.2 that a radioisotope is a form of an element with an unstable nucleus. Atoms of a radioisotope become atoms of other elements—daughter elements—as their nucleus disintegrates. This radioactive decay is not influenced by temperature, pressure, chemical bonding state, or moisture; it is influenced only by time. Thus, like the ticking of a perfect clock, each type of radioisotope decays at a constant rate. The time it takes for half of the atoms in a sample of radioisotope to decay is called a **half-life** (**FIGURE 16.12**).

Half-life is a characteristic of each radioisotope. For example, radioactive uranium 238 decays into thorium 234, which decays into something else, and so on until it becomes lead 206. The half-life of the decay of uranium 238 to lead 206 is 4.5 billion years.

The predictability of radioactive decay can be used to find the age of a volcanic rock (the date it solidified). Rock forms from magma, which is a hot, molten material deep under Earth's surface. Because magma is fluid, atoms swirl and mix in it. When the material cools, for example after reaching the surface as lava, it hardens and becomes rock. Minerals crystallize in the rock as it hardens. Each kind of mineral has a characteristic structure and composition. For example, the mineral called zircon (left) consists mainly of orderly arrays of zirconium silicate molecules ($ZrSiO_4$). Some of the molecules in a newly formed zircon crystal have uranium atoms substituted for zirconium atoms, but never lead atoms. However, uranium decays into lead at a predictable rate. Thus, over time, uranium atoms disappear from a zircon crystal, and lead atoms accumulate in it. The ratio of uranium atoms to lead atoms in a zircon crystal can be measured precisely. That ratio can be used to calculate how long ago the crystal formed (its age).

zircon

We have just described **radiometric dating**, a method that can reveal the age of a material by measuring its content and proportions of a radioisotope and daughter elements. The oldest known terrestrial rock, a tiny zircon crystal from the Jack Hills in Western Australia, formed 4.4 billion years ago. More recent fossils that still contain carbon can be radiometrically dated by measuring their content of carbon 14 (**FIGURE 16.13**). Most of the ^{14}C in a fossil will have decayed after about 60,000 years. The age of fossils older than that can be estimated by dating volcanic rock in lava flows above and below the fossil-containing layer.

CREDITS: (12, 13B, C) © Cengage Learning; (13A) © PhotoDisc/Getty Images; (in text) Courtesy of Stan Celestian/ Glendale Comunity College Earth Science Image Archive.

A *Elomeryx*, a small terrestrial animal that lived about 30 million years ago. This is a member of the same artiodactyl group (even-toed hooved mammals) that gave rise to modern representatives, including hippopotamuses. *Elomeryx* is thought to resemble a 60-million-year-old ancestor that it shares with whales.

B *Rhodhocetus kasrani*, an ancient whale that lived about 47 million years ago. Its distinctive ankle bones are evidence of a close evolutionary connection to artiodactyls. Artiodactyls are defined by the unique "double-pulley" shape of the bone (right) that forms the lower part of their ankle joint.

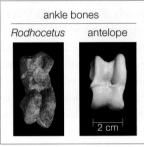

ankle bones

Rhodhocetus   antelope

2 cm

C *Dorudon atrox*, an ancient whale that lived about 37 million years ago. Its tiny, artiodactyl-like ankle bones were much too small to have supported the weight of its huge body on land, so this mammal had to be fully aquatic.

2 m

D Modern cetaceans such as the sperm whale have remnants of a pelvis and leg, but no ankle bones.

FIGURE 16.14 Comparison of cetacean skeletal features. The ancestor of whales was an artiodactyl that walked on land. Over millions of years, the lineage transitioned from life on land to life in water, and as it did, bones of the hindlimb (highlighted in blue) became smaller.

Rhodhocetus and *Dorudon* are long-lost relatives of modern whales, not direct ancestors. Both were offshoots of the ancient-artiodactyl-to-modern-whale lineage during its transition from life on land to life in water.

---

half-life Characteristic time it takes for half of a quantity of a radioisotope to decay.
radiometric dating Method of estimating the age of a rock or a fossil based on the predictability of radioactive decay.

Missing Links

The discovery of intermediate forms of cetaceans (an order of animals that includes whales, dolphins, and porpoises) offers an example of how fossil finds and radiometric dating can be used to reconstruct evolutionary history. For some time, evolutionary biologists had thought that the ancestors of modern cetaceans walked on land, then took up life in the water. Evidence in support of this idea includes a set of distinctive features of the skull and lower jaw that cetaceans share with some kinds of ancient carnivorous land animals. DNA sequence comparisons indicate that the ancient land animals were probably artiodactyls, hooved mammals with an even number of toes (two or four) on each foot (**FIGURE 16.14A**). Modern representatives of the artiodactyl lineage include camels, hippopotamuses, pigs, deer, sheep, and cows.

Until recently, we had no fossils demonstrating gradual changes in skeletal features that would have accompanied a transition of whale lineages from terrestrial to aquatic life. Researchers knew there were intermediate forms because they had found a representative fossil skull of an ancient whalelike animal, but without a complete skeleton the rest of the story remained speculative. Then, in 2000, Philip Gingerich and his colleagues unearthed complete fossilized skeletons of two ancient whales: *Rhodhocetus kasrani* (**FIGURE 16.14B**) excavated from a 47-million-year-old rock formation in Pakistan, and *Dorudon atrox* (**FIGURE 16.14C**), from 37-million-year-old rock in Egypt. Both fossil skeletons had whalelike skull bones, as well as intact ankle bones. The ankle bones of both fossils have distinctive features in common with those of extinct and modern artiodactyls. Modern cetaceans do not have even a remnant of an ankle bone (**FIGURE 16.14D**).

The proportions of limbs, skull, neck, and thorax indicate *Rhodhocetus* swam with its feet, not its tail. Like modern whales, the 5-meter (16-foot) *Dorudon* was clearly a fully aquatic tail-swimmer: The entire hindlimb was only about 12 centimeters (5 inches) long, much too small to have supported the animal's tremendous body out of water.

CREDITS: (14A, B top, C, D) © Cengage Learning; (14B left and right) © Phillip Gingerich, University of Michigan.

✔ Over billions of years, movements of Earth's outer layer of rock have changed the land, atmosphere, and oceans.

Wind, water, and other forces continuously sculpt Earth's surface, but they are only part of a much bigger picture of geological change. For example, all continents that exist today were once part of one huge supercontinent—**Pangea**—that split into fragments and drifted apart. The idea that continents move around, originally called continental drift, was proposed in the early 1900s to explain why the Atlantic coasts of South America and Africa seem to "fit" like jigsaw puzzle pieces, and why the same types of fossils occur in identical rock formations on both sides of the Atlantic Ocean. It also explained why the magnetic poles of gigantic rock formations point in different directions on different continents. As magma solidifies into rock, some iron-rich minerals in it become magnetic, and

their magnetic poles align with Earth's poles when they do. If the continents never moved, then all of these ancient rocky magnets should be aligned north-to-south, like compass needles. Indeed, the magnetic poles of rocks in each formation are aligned with one another, but the alignment is not always north-to-south. Either Earth's magnetic poles veer dramatically from their north–south axis, or the continents wander.

The idea that continents move was initially greeted with skepticism because there was no known mechanism capable of causing the movement. Then, in the late 1950s, deep-sea explorers found immense ridges and trenches stretching thousands of kilometers across the seafloor (**FIGURE 16.15**). The discovery led to the **plate tectonics theory**, which explains how continents move: Earth's outer layer of rock is cracked into immense plates, like a huge cracked eggshell. Magma welling up at an undersea ridge ❶ or continental rift at one edge of a plate pushes old rock at the opposite edge into a trench ❷. The movement is like that of a colossal conveyor belt that transports continents on top of it to new locations. The plates move no more than 10 centimeters (4 inches) a year—about half as fast as your toenails grow—but it is enough to carry a continent all the way around the world after 40 million years or so.

Evidence of tectonic movement is all around us, in faults ❸ and other geological features of our landscapes. For example, volcanic island chains (archipela-

The San Andreas Fault, extending 800 miles in California, marks the boundary between two tectonic plates.

FIGURE 16.15 Plate tectonics. Huge pieces of Earth's outer layer of rock slowly drift apart and collide. As these plates move, they convey continents around the globe.

❶ At oceanic ridges, plumes of magma (red) welling up from Earth's interior drive the movement of tectonic plates. New crust spreads outward as it forms on the surface, forcing adjacent tectonic plates away from the ridge and into trenches elsewhere.

❷ At trenches, the advancing edge of one plate plows under an adjacent plate and buckles it.

❸ Faults are ruptures in Earth's crust where plates meet. The diagram shows a rift fault, in which plates move apart. The photo above shows a strike-slip fault, in which two abutting plates slip against one another in opposite directions.

❹ Plumes of magma rupture a tectonic plate at what are called "hot spots." The Hawaiian Islands have been forming from magma that continues to erupt at a hot spot under the Pacific Plate. This and other tectonic plates are shown in Appendix V.

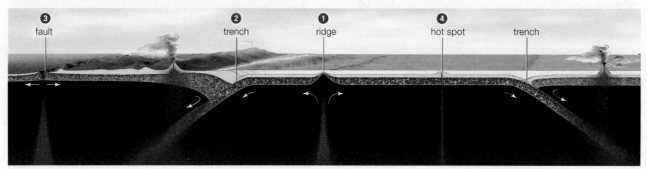

gos) form as a plate moves across an undersea hot spot. These hot spots are places where a plume of magma ruptures a tectonic plate ❹.

The fossil record also provides evidence in support of plate tectonics. Consider an unusual rock formation that exists in a huge belt across Africa. The sequence of rock layers in this formation is so complex that it is quite unlikely to have formed more than once, but identical sequences of layers also occur in huge belts that span India, South America, Madagascar, Australia, and Antarctica. Across all of these continents, the layers are the same ages. They also hold fossils found nowhere else, including imprints of the seed fern *Glossopteris* (pictured in **FIGURE 16.10C**). The most probable explanation for these observations is that the layered rock formed in one long belt on a single continent, which later broke up.

We now know that at least five times since Earth's outer layer of rock solidified 4.55 billion years ago, supercontinents formed and split up again. One called **Gondwana** formed about 500 million years ago. Over the next 230 million years, this supercontinent wandered across the South Pole, then drifted north until it merged with other landmasses to form Pangea (**FIGURE 16.16**). Most of the landmasses currently in the Southern Hemisphere as well as India and Arabia were once part of Gondwana. Some modern species, including the birds pictured in **FIGURE 16.2**, live only in these places.

Geologic changes brought on by plate tectonics have had a profound impact on life. For example, colliding continents have physically separated organisms living in oceans, and brought together those that had been living apart on land. As continents broke up, they separated organisms living on land, and brought together ones that had been living in separate oceans. These events have been a major driving force of evolution, as you will see in the next chapter.

Gondwana Supercontinent that existed before Pangea, more than 500 million years ago.
Pangea Supercontinent that formed about 270 million years ago.
plate tectonics theory Theory that Earth's outer layer of rock is cracked into plates, the slow movement of which conveys continents to new locations over geologic time.

TAKE-HOME MESSAGE 16.7
How has Earth changed over geologic time?

✔ Over geologic time, movements of Earth's crust have caused dramatic changes in continents and oceans. These changes profoundly influenced the course of life's evolution.

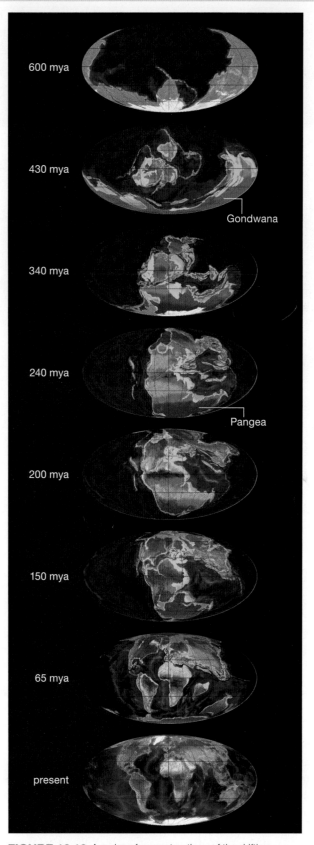

FIGURE 16.16 A series of reconstructions of the drifting continents. mya: million years ago.

16.8 Putting Time Into Perspective

Eon	Era	Period	Epoch	mya	Major Geologic and Biological Events
Phanerozoic	Cenozoic	Quaternary	Holocene	0.01	Modern humans evolve. Major extinction event is now under way.
			Pleistocene	2.6	
		Neogene	Pliocene	5.3	Tropics, subtropics extend poleward. Climate cools; dry woodlands and grasslands emerge. Adaptive radiations of mammals, insects, birds.
			Miocene	23.0	
		Paleogene	Oligocene	33.9	
			Eocene	56.0	
			Paleocene	66.0 ◄	Major extinction event
	Mesozoic	Cretaceous	Upper		Flowering plants diversify; sharks evolve. All dinosaurs and many marine organisms disappear at the end of this epoch.
				100.5	
			Lower		Climate very warm. Dinosaurs continue to dominate. Important modern insect groups appear (bees, butterflies, termites, ants, and herbivorous insects including aphids and grasshoppers). Flowering plants originate and become dominant land plants.
				145.0	
		Jurassic			Age of dinosaurs. Lush vegetation; abundant gymnosperms and ferns. Birds appear. Pangea breaks up.
				201.3 ◄	Major extinction event
		Triassic			Recovery from the major extinction at end of Permian. Many new groups appear, including turtles, dinosaurs, pterosaurs, and mammals.
				252 ◄	Major extinction event
	Paleozoic	Permian			Supercontinent Pangea and world ocean form. Adaptive radiation of conifers. Cycads and ginkgos appear. Relatively dry climate leads to drought-adapted gymnosperms and insects such as beetles and flies.
				299	
		Carboniferous			High atmospheric oxygen level fosters giant arthropods. Spore-releasing plants dominate. Age of great lycophyte trees; vast coal forests form. Ears evolve in amphibians; penises evolve in early reptiles (vaginas evolve later, in mammals only).
				359 ◄	Major extinction event
		Devonian			Land tetrapods appear. Explosion of plant diversity leads to tree forms, forests, and many new plant groups including lycophytes, ferns with complex leaves, seed plants.
				419	
		Silurian			Radiations of marine invertebrates. First appearances of land fungi, vascular plants, bony fishes, and perhaps terrestrial animals (millipedes, spiders).
				443 ◄	Major extinction event
		Ordovician			Major period for first appearances. The first land plants, fishes, and reef-forming corals appear. Gondwana moves toward the South Pole and becomes frigid.
				485	
		Cambrian			Earth thaws. Explosion of animal diversity. Most major groups of animals appear (in the oceans). Trilobites and shelled organisms evolve.
				541	
Precambrian	Proterozoic				Oxygen accumulates in atmosphere. Origin of aerobic metabolism. Origin of eukaryotic cells, then protists, fungi, plants, animals. Evidence that Earth mostly freezes over in a series of global ice ages between 750 and 600 mya.
				2,500	
	Archean and earlier				3,800–2,500 mya. Origin of bacteria and archaea.
					4,600–3,800 mya. Origin of Earth's crust, first atmosphere, first seas. Chemical, molecular evolution leads to origin of life (from protocells to anaerobic single cells).

FIGURE 16.17 The geologic time scale (above) correlated with sedimentary rock exposed by erosion in the Grand Canyon (opposite). Red triangles mark times of great mass extinctions. "First appearance" refers to appearance in the fossil record, not necessarily the first appearance on Earth. mya: million years ago. Dates are from the International Commission on Stratigraphy, 2014.

Similar sequences of sedimentary rock layers occur around the world. Transitions between the layers mark boundaries between great intervals of time in the **geologic time scale**, which is a chronology of Earth's history (**FIGURE 16.17**). Each layer's composition offers clues about conditions on Earth during the time the layer was deposited. Fossils in the layers are a record of life during that period of time.

geologic time scale Chronology of Earth's history.

TAKE-HOME MESSAGE 16.8
What is the geologic time scale?
✔ The geologic time scale correlates geological and evolutionary events of the ancient past.

Permian

Ka) Kaibab Limestone

To) Toroweap Formation

Co) Coconino Sandstone

Carboniferous

He) Hermit Shale

Es) Esplanade Sandstone

We) Wescogame Formation

Ma) Manakacha Formation

Wa) Watahomigi Formation

Re) Redwall Limestone

Te) Temple Butte Formation

Mu) Muav Limestone

Cambrian

Br) Bright Angel Shale

Ta) Tapeats Sandstone

Proterozoic

Chuar Group

Nankoweap Formation

*alUnkar Group

Vi) Vishnu Basement Rocks

*Layers not visible in this view of the Grand Canyon

Each rock layer has a composition and set of fossils that reflect events during its deposition. For example, Coconino Sandstone, which stretches from California to Montana, is mainly weathered sand. Ripple marks and reptile tracks are the only fossils in it. Many think it is the remains of a vast sand desert, similar to the modern Sahara.

FIGURE IT OUT: Which formation is marked with the (?)?   Answer: Tapeats sandstone

Reflections of a Distant Past (revisited)

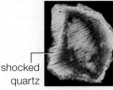

shocked quartz

The K–Pg boundary sequence is unusually rich in iridium, an element rare on Earth's surface but common in asteroids. After researchers discovered the iridium, they looked for evidence of an asteroid impact massive enough to cover the entire Earth with extraterrestrial debris. In the Yucatán Peninsula,

they found a crater so big that no one had realized it was a crater before. The K–Pg boundary sequence also contains shocked quartz and small glass spheres called tektites—rocks that form when quartz or sand (respectively) undergoes a sudden, violent application of extreme pressure. As far as we know, the only processes on Earth that produce shocked quartz and tektites are atomic bomb explosions and meteorite impacts.

summary

Section 16.1 Events of the ancient past can be explained by the same physical, chemical, and biological processes that operate today. An asteroid impact may have caused a mass extinction 66 million years ago.

Section 16.2 Expeditions by nineteenth-century naturalists yielded increasingly detailed observations of nature. Geology, **biogeography**, and **comparative morphology** of organisms and their **fossils** led to new ways of thinking about the natural world.

Section 16.3 Prevailing belief systems often influence interpretation of the cause of natural events. Nineteenth-century European naturalists proposed **catastrophism** and the **theory of uniformity** in their attempts to reconcile traditional beliefs with physical evidence of **evolution**, or change in a **lineage** over time.

Section 16.4 Humans select desirable traits in animals by selective breeding. Charles Darwin and Alfred Wallace independently came up with a theory of how environments also select traits, stated here in modern terms: A population tends to grow until it exhausts environmental resources. As that happens, competition for those resources intensifies among the population's members. Individuals with forms of shared, heritable traits that give them an advantage in this competition tend to produce more offspring. Thus, **adaptive traits (adaptations)** that impart greater **fitness** to an individual become more common in a population over generations. The process in which environmental pressures result in the differential survival and reproduction of individuals of a population is called **natural selection**. It is one of the processes that drives evolution.

Section 16.5 Fossils are typically found in stacked layers of sedimentary rock. Younger fossils usually occur in layers deposited more recently, on top of older fossils in older layers. Fossils of many organisms are relatively scarce, so the fossil record will always be incomplete.

Section 16.6 A radioisotope's characteristic **half-life** can be used to determine the age of rocks and fossils. This technique, **radiometric dating**, helps us understand the ancient history of many lineages.

Section 16.7 According to the **plate tectonics theory**, Earth's crust is cracked into giant plates that convey landmasses to new positions as they move. Earth's landmasses have periodically converged as supercontinents such as **Gondwana** and **Pangea**.

Section 16.8 Transitions in the fossil record are the boundaries of great intervals of the **geologic time scale**, a chronology of Earth's history that correlates geologic and evolutionary events.

self-quiz

Answers in Appendix VII

1. The number of species on an island depends on the size of the island and its distance from a mainland. This statement would most likely be made by _____ .
    a. an explorer   c. a geologist
    b. a biogeographer   d. a philosopher

2. The bones of a bird's wing are similar to the bones in a bat's wing. This observation is an example of _____ .
    a. uniformity   c. comparative morphology
    b. evolution   d. a lineage

3. Evolution _____ .
    a. is natural selection
    b. is change in a line of descent
    c. can occur by natural selection
    d. b and c are correct

4. A trait is adaptive if it _____ .
    a. arises by mutation   c. is passed to offspring
    b. increases fitness   d. occurs in fossils

5. In which type of rock are you most likely to find a fossil?
    a. basalt, a dark, fine-grained volcanic rock
    b. limestone, composed of sedimented calcium carbonate
    c. slate, a volcanically melted and cooled shale
    d. granite, which forms by crystallization of magma cooling below Earth's surface

Discovery of Iridium in the K–Pg Boundary Sequence In the late 1970s, geologist Walter Alvarez was investigating the composition of the K–Pg boundary sequence in different parts of the world. He asked his father, Nobel Prize–winning physicist Luis Alvarez, to help him analyze the elemental composition of the layer. The Alvarezes and their colleagues tested the K–Pg boundary sequence in Italy and Denmark. They discovered that it contains a much higher iridium content than the surrounding rock layers. Some of their results are shown in **FIGURE 16.18**.

Sample Depth	Average Abundance of Iridium (ppb)
+ 2.7 m	< 0.3
+ 1.2 m	< 0.3
+ 0.7 m	0.36
boundary layer	41.6
− 0.5 m	0.25
− 5.4 m	0.30

Iridium belongs to a group of elements (Appendix I) that are much more abundant in asteroids and other solar system materials than they are in Earth's crust. The Alvarez group concluded that the K–Pg boundary sequence must have originated with extraterrestrial material. They calculated that an asteroid 14 kilometers (8.7 miles) in diameter would contain enough iridium to account for the extra iridium in the K–Pg boundary sequence.

1. What was the iridium content of the K–Pg boundary sequence?

2. How much higher was the iridium content of the boundary sequence than the sample taken 0.7 meter above the sequence?

FIGURE 16.18 Abundance of iridium in and near the K–Pg boundary sequence in Stevns Klint, Denmark. Many rock samples taken from above, below, and at the boundary were tested for iridium content. Depths are given as meters above or below the boundary.

The iridium content of an average Earth rock is 0.4 parts per billion (ppb) of iridium. An average meteorite contains about 550 parts per billion of iridium.

The photo shows Luis and Walter Alvarez with a section of the boundary sequence.

---

6. Which of the following is a fossil?
   a. an insect encased in 10-million-year-old tree sap
   b. a woolly mammoth frozen in Arctic permafrost for the last 50,000 years
   c. mineral-hardened remains of a whalelike animal found in an Egyptian desert
   d. an impression of a plant leaf in a rock
   e. all of the above can be considered fossils

7. If the half-life of a radioisotope is 20,000 years, then a sample in which three-quarters of that radioisotope has decayed is _____ years old.
   a. 15,000   b. 26,667   c. 30,000   d. 40,000

8. Did Pangea or Gondwana form first?

9. The dinosaurs died out ___66___ million years ago.

10. On the geologic time scale, life originated in the _____ .
    a. Archean   c. Phanerozoic
    b. Proterozoic   d. Cambrian

11. Match the terms with the most suitable description.
    __a__ fitness   a. measured by reproductive success
    __d__ fossils   b. geologic change occurs
    __e__ natural   continuously
         selection   c. geologic change occurs
    __f__ half-life   in unusual major events
    __c__ catastrophism   d. evidence of life in distant past
    __b__ uniformity   e. survival of the fittest
                          f. characteristic of a radioisotope

12. Forces that cause geologic change include _____ (select all that are correct).
    a. erosion   d. tectonic plate movement
    b. natural selection   e. wind
    c. volcanic activity   f. meteorite impacts

CENGAGE To access course materials, please visit
brain.com www.cengagebrain.com.

critical thinking

1. Radiometric dating does not measure the age of an individual atom. It is a measure of the age of a quantity of atoms— a statistic. As with any statistical measure, its values may deviate around an average (see sampling error, Section 1.8). Imagine that one sample of rock is dated ten different ways. Nine of the tests yield an age close to 225,000 years. One test yields an age of 3.2 million years. Do the nine consistent results imply that the one that deviates is incorrect, or does the one odd result invalidate the nine that are consistent?

2. If you think of geologic time spans as minutes, life's history might be plotted on a clock such as the one shown below. According to this clock, the most recent epoch started in the last 0.1 second before noon. Where does that put you?

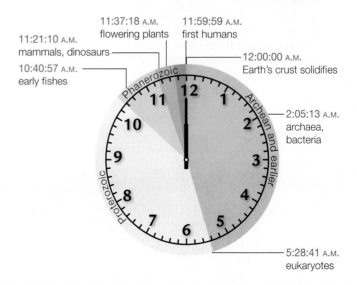

11:37:18 A.M. flowering plants
11:59:59 A.M. first humans
11:21:10 A.M. mammals, dinosaurs
12:00:00 A.M. Earth's crust solidifies
10:40:57 A.M. early fishes
Phanerozoic
Archean and earlier
2:05:13 A.M. archaea, bacteria
Proterozoic
5:28:41 A.M. eukaryotes

CREDITS: (18) left, Lawrence Berkeley National Laboratory; right, © Cengage Learning; (CT #2) © Cengage Learning.

MICROEVOLUTION

Members of a population inherit different alleles, which are the basis of differences in phenotype. An allele may increase or decrease in frequency in a population, a change called microevolution.

PATTERNS OF NATURAL SELECTION

Natural selection is one of the processes that drive microevolution. Depending on the population and its environment, natural selection can shift or maintain a range of variation in a heritable trait.

OTHER PROCESSES OF MICROEVOLUTION

With genetic drift, change can occur in a line of descent by chance alone. Gene flow counters the evolutionary effects of mutation, natural selection, and genetic drift.

HOW SPECIES ARISE

Speciation typically starts after gene flow ends. Microevolution leads to genetic divergences, which are reinforced as mechanisms evolve that prevent interbreeding.

MACROEVOLUTION

Macroevolutionary patterns include the origin of major groups, one species giving rise to many, two species evolving jointly, and mass extinctions.

Chapter 18 explores the techniques we use to keep track of species and evolutionary patterns. Later chapters return to polyploid plants (Section 29.9), the genetic basis of behavior (Section 43.2), mating behaviors (Section 43.7), examples of natural selection at work in populations (Section 44.6), coevolved species (45.6), and competition and other interactions between species (Chapter 45).

Scarlet fever, tuberculosis, and pneumonia once caused one-fourth of the annual deaths in the United States. Since the 1940s, we have been relying on antibiotics to fight these and other dangerous bacterial diseases. We have also been using them in other, less dire circumstances. For an as-yet-unknown reason, antibiotics promote growth in cattle, pigs, poultry, and even fish. The agricultural industry uses a lot of antibiotics, mainly for this purpose. In 2011, 13.7 million kilograms (about 30 million pounds) of antibiotics were used for agriculture in the U.S.—more than four times the amount used to treat people in the same year.

A natural population of bacteria is diverse, and it can evolve astonishingly fast. Consider how each cell division is an opportunity for mutation (Section 8.6). The common intestinal bacteria *E. coli* can divide every 17 minutes, so even if a population starts out as clones, its cells diversify quickly. In addition, bacteria share DNA even among different species, and this adds even more genetic diversity to their populations.

Genetic diversity is an advantage in a changing environment (Section 12.1). When a natural population of bacteria is exposed to a selection pressure such as an antibiotic, some cells in the population are likely to survive because they carry an allele that offers an advantage—antibiotic resistance, in this case. As susceptible cells die and the survivors reproduce, the frequency of antibiotic-resistance alleles in the population increases. A typical two-week course of treatment with antibiotics can exert selection pressure on over a thousand generations of bacteria. The pressure drives genetic change in bacterial populations so they become composed mainly of antibiotic resistant cells. Thus, the practice of treating livestock with growth-promoting antibiotics essentially guarantees the production of antibiotic-resistant bacterial populations (**FIGURE 17.1**).

Farms where antibiotics are used to promote growth are hot spots for the spread of resistant bacteria to humans. Veterinarians and other people who work with the animals on these farms tend to carry more antibiotic-resistant bacteria in their bodies. So do neighbors who live within a mile. The bacteria spread much farther than the farm, however. Bacteria on an animal's skin or in its digestive tract can easily contaminate its meat during slaughter (Section 4.1), and contaminated meat ends up in restaurant and home kitchens. A 2013 investigation found "worrisome" amounts of bacteria in 97% of the chicken meat in stores across the United States. About half of the samples tested were contaminated with superbugs—bacteria that are resistant to

FIGURE 17.1 The vast majority of chickens raised for meat in the United States spend their lives in gigantic flocks that crowd huge buildings like this one. Growth-promoting antibiotics are given to the entire flock in food, a practice that pressures normal bacterial populations to become antibiotic resistant.

multiple antibiotics—and one in ten contained multiple superbug species. An earlier study found antibiotic-resistant bacteria in more than half of supermarket ground beef and pork chops, and in over 80 percent of ground turkey. Bacteria can be killed by the heat of cooking, but it is almost impossible to prevent them from spreading from contaminated meat to kitchen surfaces—and to people—during the process.

We have only a limited number of antibiotic drugs, and developing new ones is much slower than bacterial evolution. As resistant bacteria become more common, the number of antibiotics that can be used to effectively treat infections in humans dwindles. Using a particular antibiotic only in animals, or only in humans, is not a solution to this problem; there are only a few mechanisms by which these drugs kill bacteria, so resistance to one antibiotic often confers resistance to others. For example, bacteria that are resistant to flavomycin (a phosphoglycolipid antibiotic used only in animals) also resist vancomycin (a glycopeptide antibiotic used only in humans). Superbugs resistant to most currently available antibiotics are turning up at an alarming rate.

All of this amounts to bad news. An infection with antibiotic-resistant bacteria tends to be longer, more severe, and more likely to be deadly than one more easily treatable with antibiotics. Superbugs cause more than 2 million cases of serious illness each year in the United states alone; they outright kill 23,000 of these people. Many, many more die because the infection complicates another, preexisting illness.

✔ Mutations in individuals are the original source of new alleles in a population's pool of genetic resources.

✔ A change in an allele's frequency in a population is called microevolution.

Alleles in Populations

Section 1.2 introduced a **population** as a group of interbreeding individuals of the same species in some specified area. The individuals of a population (and a species) share certain features. For example, giraffes normally have long necks, brown spots on white coats, and so on. These are examples of morphological traits (*morpho*– means form). Individuals of a species also share physiological traits, such as metabolic activities. They also respond the same way to certain stimuli, as when hungry giraffes feed on tree leaves. These are behavioral traits.

Members of a population have the same traits because they have the same genes. However, almost every shared trait varies a bit among individuals of a population (**FIGURE 17.2**). Alleles of the shared genes are the basis of this variation. Many traits have two or more distinct forms, or morphs. A trait with only two forms is dimorphic (*di*– means two). Purple and white flower color in the pea plants that Gregor Mendel studied is an example of a dimorphic trait (Section 13.3). Dimorphic flower color occurs in this case because the interaction of two alleles with a clear dominance relationship gives rise to the trait. Traits with more than two distinct forms are polymorphic (*poly*–, many). ABO blood type in humans, which is determined by the codominant alleles of the *ABO* gene, is an example (Section 13.5). The genetic basis of traits that vary continuously among the individuals of a population is typically quite complex (Sections 13.5–13.7). Any or all of the genes that influence such traits may have multiple alleles.

In earlier chapters, you learned about genetic events that contribute to the variation in shared traits we see among individuals of a population (**TABLE 17.1**). Mutation is the original source of new alleles. Other events shuffle alleles into different combinations, and what a shuffle that is! There are $10^{116,446,000}$ possible combinations of human alleles. Not even 10^{10} people are living today. Unless you have an identical twin, it is extremely unlikely that another person with your precise genetic makeup has ever lived, or ever will.

An Evolutionary View of Mutations

Being the original source of new alleles, mutations are worth another look, this time in the context of their

Table 17.1 Some Sources of Variation in Shared Traits

Genetic Event	Effect
Mutation	Original source of new alleles
Crossing over at meiosis I	Introduces new combinations of alleles into chromosomes
Independent assortment at meiosis I	Mixes maternal and paternal chromosomes
Fertilization	Combines alleles from two parents
Changes in chromosome number or structure	Often dramatic changes in structure and function

impact on populations. We cannot predict when or in which individual a particular gene will mutate. We can, however, predict the average mutation rate of a species, which is the probability that a mutation will occur in a given interval. In the human species, that rate is about 2.2×10^{-9} mutations per base pair per year. In other words, about 70 nucleotides in the human genome sequence change every decade.

In humans at least, most mutations are neutral. A **neutral mutation** changes the DNA sequence of a chromosome, but the alteration has no effect on survival or reproduction—it neither helps nor hurts the individual. For instance, if you carry a mutation that keeps your earlobes attached to your head instead of swinging freely, attached earlobes should not in itself stop you from surviving and reproducing as well as anybody else. So, natural selection would not affect the frequency of this mutation in the human population.

Some mutations give rise to structural, functional, or behavioral alterations that reduce an individual's chances of surviving and reproducing. Even one biochemical change may be devastating. For instance, the skin, bones, tendons, lungs, blood vessels, and other vertebrate organs incorporate the protein collagen. If one of the genes for collagen mutates in a way that changes the protein's function, the entire body may be affected. A mutation such as this can change phenotype so drastically that it results in death, in which case it is a **lethal mutation**.

Occasionally, a change in the environment favors a mutation that had previously been neutral or even somewhat harmful. Even if a beneficial mutation bestows only a slight advantage, its frequency tends to increase in a population over time. This is because

FIGURE 17.2 Sampling morphological variation among zigzag Nerite snails (left) and humans (right). Variation in shared traits among individuals is mainly an outcome of variations in alleles that influence those traits.

natural selection operates on traits with a genetic basis. With natural selection, remember, environmental pressures result in an increase in the frequency of an adaptive form of a trait in a population over generations (Section 16.4). Mutations have been altering genomes for billions of years, and they continue to do so. Cumulatively, mutations have given rise to Earth's staggering biodiversity. Think about it: The reason you do not look like an avocado or an earthworm or even your next-door neighbor began with mutations that occurred in different lines of descent.

Allele Frequencies

Together, all the alleles of all the genes of a population make up a pool of genetic resources—a **gene pool**. Members of a population breed with one another more often than they breed with members of other popula-

tions, so their gene pool is more or less isolated. **Allele frequency** refers to the abundance of a particular allele among the individuals of a population, expressed as a fraction of the total number of alleles. Any change in an allele's frequency in the gene pool of a population (or a species) is called **microevolution**.

Microevolution is always occurring in natural populations because, as you will see in the next section, processes that drive it are always operating in nature. The remaining sections of this chapter explore some of the processes and evolutionary effects of mutation, natural selection, genetic drift, and gene flow. As you learn about these patterns, remember an important point: Evolution is not purposeful; it simply fills the nooks and crannies of opportunity.

allele frequency Abundance of a particular allele among members of a population.
gene pool All the alleles of all the genes in a population; a pool of genetic resources.
lethal mutation Mutation that alters phenotype so drastically that it causes death.
microevolution Change in an allele's frequency in a population.
neutral mutation A mutation that has no effect on survival or reproduction.
population A group of organisms of the same species who live in a specific location and breed with one another more often than they breed with members of other populations.

TAKE-HOME MESSAGE 17.2
What is microevolution?

✔ Individuals of a natural population share morphological, physiological, and behavioral traits characteristic of the species. Alleles are the main basis of differences in the details of those shared traits.

✔ All alleles of all individuals in a population make up the population's gene pool. An allele's abundance in the gene pool is called its allele frequency.

✔ Microevolution is change in allele frequency. It is always occurring in natural populations because processes that drive it are always operating.

✔ Natural populations are always evolving.

✔ Researchers trace evolution within a population by tracking deviations from a baseline of genetic equilibrium.

Early in the twentieth century, Godfrey Hardy (a mathematician) and Wilhelm Weinberg (a physician) independently applied the rules of probability to population genetics. Both realized that, under certain theoretical conditions, allele frequencies in a sexually reproducing population's gene pool would remain stable from one generation to the next. The population would stay in

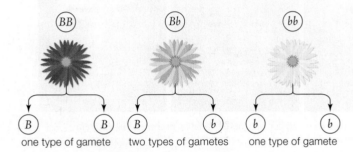

A In this two-allele system, *B* specifies dark blue flowers; *b*, white. Plants that are homozygous (*BB* or *bb*) make one kind of gamete. Heterozygous plants (*Bb*) have light blue flowers and make two kinds of gametes (*B* and *b*).

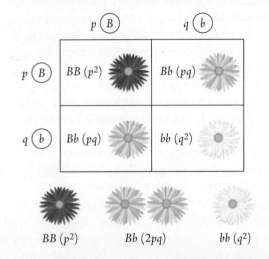

B Say *p* is the proportion of *B* alleles in the gene pool, and *q* is the proportion of *b* alleles. This Punnett square shows that in each generation of a randomly mating population, the predicted proportion of offspring that will inherit two *B* alleles is $p \times p$, or p^2. Likewise, the proportion that will inherit both alleles is $2pq$, and the proportion that will inherit two *b* alleles is q^2.

FIGURE 17.3 ▶**Animated** Hardy–Weinberg calculations. In this example, two alleles show incomplete dominance over flower color.

FIGURE IT OUT If 1/4 of this population has dark blue flowers and 1/4 has white flowers, what proportion of the next generation will have light blue flowers (assuming genetic equilibrium)?

Answer: Half of the gametes have a *B* allele; the other half have a *b* allele. Both *p* and *q* = 0.5, so 2*pq* = 50 percent.

this stable state, called **genetic equilibrium**, as long as all of the following five conditions are met:

1. Mutations never occur.
2. The population is infinitely large.
3. The population is isolated from all other populations (no individual enters or leaves).
4. Mating is random.
5. All individuals survive and produce the same number of offspring.

As you can imagine, all five of these conditions never occur in nature, so natural populations are never in genetic equilibrium.

Applying the Hardy–Weinberg Law

The concept of genetic equilibrium under ideal conditions is called the Hardy–Weinberg law. To see how it works, consider a hypothetical gene that encodes a blue pigment in daisies. A plant homozygous for one allele (*BB*) has dark blue flowers. A plant homozygous for the other allele (*bb*) has white flowers. These two alleles are inherited in a pattern of incomplete dominance, so a heterozygous plant (*Bb*) has medium-blue flowers (**FIGURE 17.3A**).

Start with the concept that allele frequencies always add up to one. For a gene with two alleles, the following equation is true:

$$p + q = 1.0$$

where *p* is the frequency of one allele in the population, and *q* is the frequency of the other. These alleles assort into different gametes during meiosis (Section 13.3), and then meet up at fertilization. If our hypothetical population of plants mate at random, the fraction of offspring that inherit two *B* alleles (*BB*) is $p \times p$, or p^2; the fraction that inherit two *b* alleles (*bb*) is q^2; and the fraction that inherit one *B* allele and one *b* allele (*Bb*) is $2pq$ (**FIGURE 17.3B**). Note that the frequencies of the three genotypes, whatever they may be, add up to 1.0:

$$p^2 + 2pq + q^2 = 1.0$$

Imagine that a population of daisies consists of 1,000 plants: 490 homozygous (*BB*), 420 heterozygous (*Bb*), and 90 homozygous (*bb*), and each of these individuals makes just two gametes. The *BB* individuals make 980 gametes, all with the *B* allele. The *Bb* individuals make 840 gametes, half (420) with the *B* allele. Thus, the frequency of the *B* allele among the pool of gametes is:

$$B\ (p) = \frac{980 + 420}{2,000\ \text{alleles}} = \frac{1,400}{2,000} = 0.7$$

The *bb* individuals make 180 gametes, all with the *b* allele. The other half of the 840 gametes made by the heterozygous individuals also have the *b* allele. Thus, the frequency of the *b* allele among the population's pool of gametes is:

$$b\ (q)\ =\ \frac{180\ +\ 420}{2{,}000\ \text{alleles}}\ =\ \frac{600}{2{,}000}\ =\ 0.3$$

With $p = 0.7$ and $q = 0.3$, the proportion of genotypes among the next generation of individuals should be:

BB	(p^2)	=	$(0.7)^2$	=	0.49
Bb	($2pq$)	=	$2\ (0.7 \times 0.3)$	=	0.42
bb	(q^2)	=	$(0.3)^2$	=	0.09

These proportions are the same as the ones in the parent population. As long as the five conditions required for genetic equilibrium are met, traits specified by the alleles should show up in the same proportions in each generation. If they do not, the population is evolving.

Real-World Situations

Genetic equilibrium is often used as a benchmark. For example, researchers used it to determine the carrier frequency of an allele that causes a genetic disorder called hereditary hemochromatosis (HH). Individuals affected by HH absorb too much iron from their food, and this causes liver problems, fatigue, and arthritis. The allele is inherited in an autosomal recessive pattern, so carriers show no symptoms. The researchers found the allele's frequency among people of Irish ancestry to be 14 percent. If $q = 0.14$, then p, the frequency of the normal allele, must be 0.86. Thus, the carrier frequency, $2pq$, was calculated to be 0.24 (24 percent of the population).

As another example, consider the *BRCA* genes, mutations in which are linked to breast cancer (Section 15.1). A deviation from predicted allele frequencies suggested that *BRCA* mutations have effects even before birth, so researchers investigated the frequency of mutated alleles of these genes among newborn girls. They found fewer individuals homozygous for these alleles than expected, based on the number of heterozygous individuals. Thus, in homozygous form, *BRCA* mutations impair the survival of female embryos.

---

genetic equilibrium Theoretical state in which an allele's frequency never changes in a population's gene pool.

17.4 Patterns of Natural Selection

✔ Natural selection occurs in different patterns depending on the organisms involved and their environment.

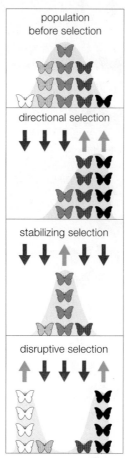

FIGURE 17.4 Overview of three modes of natural selection.

The rest of this chapter explores the mechanisms and effects of natural selection and other processes that drive evolution. Remember from Section 16.4 that natural selection is a process in which environmental pressures result in the differential survival and reproduction of individuals of a population based on their shared, heritable traits. It influences the frequency of alleles in a population by operating on traits with a genetic basis.

We observe different patterns of natural selection. In some cases, individuals with a trait at one extreme of a range of variation are selected against, and forms at the other extreme are adaptive. We call this directional selection. With stabilizing selection, midrange forms of a trait are adaptive, and extremes are selected against. With disruptive selection, forms at the extremes of the range of variation are adaptive, and intermediate forms are selected against. These modes of natural selection, which **FIGURE 17.4** summarizes, are discussed in the next two sections.

Section 17.7 explores sexual selection, a mode of natural selection that operates on a population by influencing mating success. This section also discusses balanced polymorphism, a particular case in which natural selection maintains a relatively high frequency of multiple alleles in a population. Natural selection and other processes of evolution can alter a population so much that it becomes a new species. We discuss mechanisms of speciation in the final sections.

Directional Selection

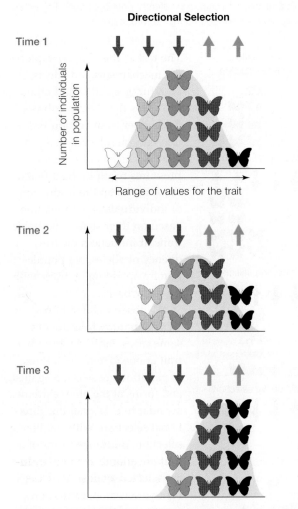

FIGURE 17.5 ▶Animated With directional selection, a form of a trait at one end of a range of variation is adaptive. Bell-shaped curves indicate continuous variation. Red arrows indicate which forms are being selected against; green, forms that are adaptive.

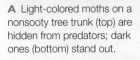

A Light-colored moths on a nonsooty tree trunk (top) are hidden from predators; dark ones (bottom) stand out.

B In places where soot darkens tree trunks, the dark color (bottom) provides more camouflage than the light color (top).

FIGURE 17.6 ▶Animated Adaptive value of two color forms of the peppered moth.

✔ Changing environmental conditions can result in a directional shift in an allele's frequency.

Directional selection shifts allele frequencies in a consistent direction, so forms at one end of a range of phenotypic variation become more common over time (**FIGURE 17.5**). Antibiotic use that fosters resistant bacterial populations is one example of directional selection. Additional examples follow.

Examples of Directional Selection

The Peppered Moth A well-documented case of directional selection involves coloration changes in peppered moths. These moths feed and mate at night, then rest on trees during the day. In preindustrial England, the vast majority of peppered moths were white with black speckles, and a small number were much darker. At this time, the air was clean, and light-gray lichens grew on the trunks and branches of most trees. When light-colored moths rested on lichen-covered trees, they were well camouflaged, whereas darker moths were not (**FIGURE 17.6A**). By the 1850s, the industrial revolution had begun, and smoke emitted by coal-burning factories was killing the lichens. Dark moths, which were better camouflaged on lichen-free, soot-darkened trees, had become more common (**FIGURE 17.6B**).

Scientists suspected that predation by birds was the selective pressure that shaped moth coloration, and in the 1950s, H. B. Kettlewell set out to test this hypothesis. He bred dark and light moths in captivity, marked them for easy identification, then released them in several areas. His team recaptured more of the dark moths in the polluted areas and more light ones in the less polluted ones. The researchers also observed predatory birds eating more light-colored moths in soot-darkened forests, and more dark-colored moths in cleaner, lichen-rich forests. Dark-colored moths were clearly at a selective advantage in industrialized areas.

Pollution controls went into effect in 1952. As a result of improved environmental standards, tree trunks gradually became free of soot, and lichens made a comeback. Kettlewell observed that moth phenotypes shifted too: Wherever pollution decreased, the frequency of dark moths decreased as well. Recent research has confirmed Kettlewell's results implicating birds as selective agents of peppered moth coloration. It has also shown that coloration in peppered moths is determined by a single gene. Individuals with a dominant allele of this gene are dark; those homozygous for a recessive allele are light.

CREDITS: (5) © Cengage Learning; (6) J. A. Bishop, L. M. Cook.

Rock Pocket Mice Directional selection also affects the color of rock pocket mice in Arizona's Sonoran Desert. Rock pocket mice are small mammals that spend the day sleeping in underground burrows, emerging at night to forage for seeds. Light brown granite dominates their environment, but there are also patches of dark basalt: the remains of ancient lava flows. Most of the mice in populations that inhabit the dark rock have dark gray coats (**FIGURE 17.7A**). Most of the mice in populations that inhabit the light brown rock have light brown coats (**FIGURE 17.7B**). The difference arises because mice that match the rock color in each habitat are camouflaged from their natural predators. Night-flying owls more easily see mice that do not match the rocks, and they preferentially eliminate easily seen mice from each population. Thus, in both habitats, selective predation has resulted in a directional shift in the frequency of alleles that affect coat color.

Warfarin Resistance in Rats Rats thrive in urban centers, where garbage is plentiful and natural predators are not. Part of their success stems from an ability to reproduce very quickly: Rat populations can expand within weeks to match the amount of garbage available for them to eat. For decades, people have been fighting back with poisons. Baits laced with warfarin, an organic compound that interferes with blood clotting, became popular in the 1950s. Rats that ate the poisoned baits died within days after bleeding internally or losing blood through cuts or scrapes. Warfarin was extremely effective, and its impact on harmless species was much lower than that of other rat poisons. It quickly became the rat poison of choice. By 1980, however, about 10 percent of rats in urban areas were resistant to warfarin. What happened?

Warfarin interferes with blood clotting because it inhibits the function of an enzyme called VKORC1. This enzyme regenerates vitamin K, which participates as a cofactor in the post-translational modification of blood clotting factors (Section 14.4). When vitamin K is not regenerated, the clotting factors are not properly processed, and clotting cannot occur. Rats resistant to warfarin have a mutated version of the *VKORC1* gene; the enzyme encoded by this allele is insensitive to warfarin.

"What happened" was evolution by natural selection. Rats with the normal allele died after eating warfarin; the lucky ones with a mutated allele survived and passed it to their offspring. The rat populations

A Mice with dark fur are better camouflaged—and more common—in areas of dark basalt rock.

B Mice with light fur are better camouflaged—and more common—in areas dominated by light-colored granite.

FIGURE 17.7 Directional selection in the rock pocket mouse (*Chaetodipus intermedius*). Predators preferentially eliminate individuals with coat colors that do not match their surroundings in the Sonoran Desert.

recovered quickly, and a higher proportion of individuals in the next generation carried a mutation. With each onslaught of warfarin, the frequency of the mutation in rat populations increased. Exposure to warfarin had exerted directional selection.

The mutation that results in warfarin resistance also reduces the activity of the VKORC1 enzyme, so rats that have it require a lot of extra vitamin K. However, being vitamin K deficient is not so bad when compared with being dead from rat poison. In the absence of warfarin, though, rats with the allele are at a serious disadvantage because they cannot easily obtain enough vitamin K from their diet to sustain normal blood clotting and bone formation. Thus, the frequency of a warfarin resistance allele in a rat population declines quickly after warfarin exposure ends.

TAKE-HOME MESSAGE 17.5

What is directional selection?

✔ With directional selection, a range of variation in a trait shifts in a consistent direction. The frequency of alleles underlying the trait also shifts directionally.

directional selection Mode of natural selection in which phenotypes at one end of a range of variation are favored.

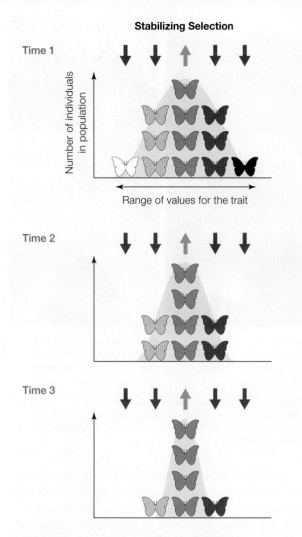

Stabilizing Selection

Time 1

Number of individuals in population

Range of values for the trait

Time 2

Time 3

FIGURE 17.8 ▶Animated With stabilizing selection, extreme forms of a trait are eliminated, and an intermediate form is maintained. Red arrows indicate which forms are being selected against; green, the form that is adaptive. Compare the data set from a field experiment in **FIGURE 17.9**.

✔ Stabilizing selection is a mode of natural selection in which an intermediate phenotype is adaptive.

✔ Disruptive selection is a mode of natural selection in which extreme forms of a trait are adaptive.

Natural selection does not always result in a directional shift in a population's range of phenotypes. In some cases, environmental pressures favor a midrange form of a trait; in others, a midrange form is eliminated and the most extreme forms are adaptive.

Stabilizing Selection

With **stabilizing selection**, an intermediate form of a trait is favored, and extreme forms are selected against (**FIGURE 17.8**).

Consider how environmental pressures maintain an intermediate body mass in populations of sociable weavers (**FIGURE 17.9**). These birds live in the African savanna, where they build large communal nests. Their body mass has a genetic basis. Between 1993 and 2000, Rita Covas and her colleagues investigated selection pressures that operate on sociable weaver body mass by capturing and weighing thousands of birds before and after the breeding seasons. The results of this study indicated that optimal body mass in sociable weavers is a trade-off between the risks of starvation and predation. Birds that carry less fat are more likely to starve than fatter birds. However, birds that carry more fat spend more time eating, which in this species means foraging in open areas where they are easily accessible to predators. Fatter birds are also more attractive to predators, and not as agile when escaping. Thus, predators are agents of selection that eliminate the fattest individuals. Birds of intermediate weight have

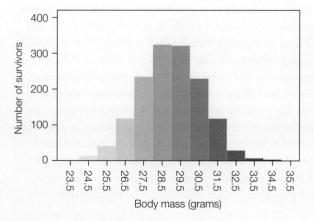

Number of survivors

Body mass (grams)

FIGURE 17.9 Stabilizing selection in sociable weavers (left). Graph (right) shows the number of birds (out of 977) that survived a breeding season. **FIGURE IT OUT** According to these data, what is the optimal weight of a sociable weaver? Answer: About 29 grams

the selective advantage, and they make up the bulk of sociable weaver populations.

Disruptive Selection

With **disruptive selection**, forms of a trait at both ends of a range of variation are favored, and intermediate forms are selected against (**FIGURE 17.10**).

The black-bellied seedcracker is a colorful finch species native to Cameroon, Africa. In these birds, there is a genetic basis for bill size. The bill of a typical black-bellied seedcracker, male or female, is either 12 millimeters wide, or wider than 15 millimeters (**FIGURE 17.11**). Birds with a bill size between 12 and 15 millimeters are uncommon. It is as if every human adult were 4 feet or 6 feet tall, with no one of intermediate height. Seedcrackers with the large and small bill forms inhabit the same geographic range, and they breed randomly with respect to bill size.

The dimorphism in bill size of seedcrackers arises from (and is maintained by) environmental factors that affect feeding performance. The finches feed mainly on the seeds of two types of sedge, which is a grass-like plant. One sedge produces hard seeds; the other produces soft seeds. Small-billed birds are better at opening the soft seeds, but large-billed birds are better at cracking the hard ones. Both hard and soft sedge seeds are abundant during Cameroon's semiannual wet seasons. At these times, all seedcrackers feed on both seed types. The seeds become scarce during the region's dry seasons. As competition for food intensifies, each bird focuses on eating the seeds that it opens most efficiently: Small-billed birds feed mainly on soft seeds, and large-billed birds feed mainly on hard seeds. Birds with intermediate-sized bills cannot open either type of seed as efficiently as the other birds, so they are less likely to survive the dry seasons.

---

disruptive selection Mode of natural selection in which traits at the extremes of a range of variation are adaptive, and intermediate forms are not.
stabilizing selection Mode of natural selection in which an intermediate form of a trait is adaptive, and extreme forms are not.

TAKE-HOME MESSAGE 17.6
In what modes of natural selection are intermediate or extreme forms of traits adaptive?

✔ With stabilizing selection, an intermediate phenotype is adaptive, and extreme forms are selected against.

✔ With disruptive selection, an intermediate form of a trait is selected against, and extreme phenotypes are adaptive.

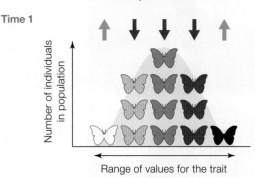

Disruptive Selection

Time 1

Number of individuals in population

Range of values for the trait

Time 2

Time 3

FIGURE 17.10 ▶**Animated** With disruptive selection, a midrange form of a trait is eliminated, and extreme forms are maintained. Red arrows indicate which form is being selected against; green, the forms that are adaptive.

lower bill 12 mm wide

lower bill 15 mm wide

FIGURE 17.11 ▶**Animated** Disruptive selection in African seedcracker populations maintains a distinct dimorphism in bill size.

Competition for scarce food during dry seasons favors birds with bills that are either 12 millimeters wide (left) or 15 to 20 millimeters wide (right). Birds with bills of intermediate size are selected against.

A Male elephant seals engaged in combat. Males of this species typically compete for access to clusters of females.

B A male bird of paradise engaged in a flashy courtship display has caught the eye (and, perhaps, the sexual interest) of a female. Female birds of paradise are choosy; a male mates with any female that accepts him.

C Female stalk-eyed flies prefer to mate with males that have the longest eyestalks, a trait that provides no known selective advantage other than sexual attractiveness.

FIGURE 17.12 Sexual selection in action.

✔ Some adaptive traits help individuals secure mates.

✔ Any mode of natural selection may maintain multiple alleles in a population.

Survival of the Sexiest

Not all evolution is driven by selection for traits that enhance survival. Competition for mates is another selective pressure that can shape form and behavior. Consider dimorphisms among males and females of some sexually reproducing species (a trait that differs among males and females is called a **sexual dimorphism**). Individuals of one sex are more colorful, larger, or more aggressive than individuals of the other sex. These traits can seem puzzling because they take energy and time away from activities that enhance survival, and some actually hinder an individual's ability to survive. Why, then, do they persist? The answer is **sexual selection**, in which the evolutionary winners outreproduce others of a population because they are better at securing mates. With this mode of natural selection, the most adaptive forms of a trait are those that help individuals defeat rivals for mates, or are most attractive to the opposite sex.

For example, the females of some species cluster in defensible groups when they are sexually receptive, and males compete for sole access to the groups. Competition for the ready-made harems favors brawny, combative males (**FIGURE 17.12A**).

Males or females that are choosy about mates act as selective agents on their own species. The females of some species shop for a mate among males that display species-specific cues such as a highly specialized appearance or courtship behavior (**FIGURE 17.12B**). The cues often include flashy body parts or movements, traits that tend to attract predators and in some cases are a physical hindrance. However, to a female member of the species, a flashy male's survival despite his obvious handicap may imply health and vigor, two traits that are likely to improve her chances of bearing healthy, vigorous offspring. Selected males pass alleles for their attractive traits to the next generation of males, and females pass alleles that influence mate preference to the next generation of females. Highly exaggerated traits can be the evolutionary outcome (**FIGURE 17.12C**).

Maintaining Multiple Alleles

Any mode of natural selection may keep two or more alleles of a gene circulating at relatively high frequency in a population's gene pool, a state called **balanced polymorphism**. For example, sexual selection maintains multiple alleles that govern eye color in populations of

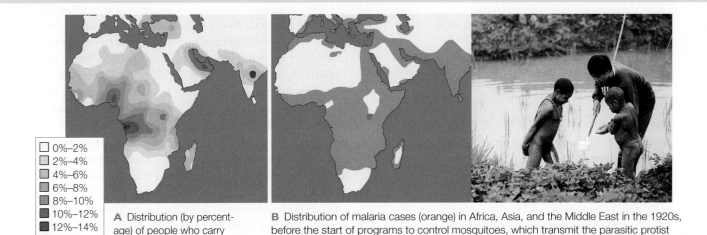

A Distribution (by percentage) of people who carry the sickle-cell allele.

- ☐ 0%–2%
- ☐ 2%–4%
- ☐ 4%–6%
- ☐ 6%–8%
- ☐ 8%–10%
- ☐ 10%–12%
- ☐ 12%–14%
- ■ >14%

B Distribution of malaria cases (orange) in Africa, Asia, and the Middle East in the 1920s, before the start of programs to control mosquitoes, which transmit the parasitic protist that causes the disease. Notice the correlation with the distribution of the sickle-cell allele in A. The photo shows a physician searching for mosquito larvae in Southeast Asia.

FIGURE 17.13 Malaria and sickle-cell anemia.

Drosophila fruit flies. Female flies prefer to mate with rare white-eyed males, until the white-eyed males become more common than red-eyed males, at which point the red-eyed flies are again preferred. This is also an example of **frequency-dependent selection**, in which the adaptive value of a particular form of a trait depends on its frequency in a population.

Balanced polymorphism can also arise in environments that favor heterozygous individuals. Consider the gene that encodes the beta globin chain of hemoglobin. *HbA* is the normal allele; the codominant *HbS* allele carries a mutation that causes sickle-cell anemia (Section 9.6). Even with medical care, about 15 percent of individuals homozygous for the *HbS* allele die by age 18 from complications of the disorder.

Despite being so harmful, the *HbS* allele persists at very high frequency among the human populations in tropical and subtropical regions of Asia, Africa, and the Middle East (**FIGURE 17.13A**). Why? Populations with the highest frequency of the *HbS* allele also have the highest incidence of malaria (**FIGURE 17.13B**). Mosquitoes transmit *Plasmodium*, the parasitic protist that causes malaria, to human hosts (more about this in Section 21.7). *Plasmodium* multiplies in the liver and then in red blood cells. The cells rupture and release new parasites during recurring bouts of severe illness.

People who make both normal and sickle hemoglobin are more likely to survive malaria than people who make only normal hemoglobin. In *HbA/HbS* heterozygous individuals, *Plasmodium*-infected red blood cells sometimes sickle. The abnormal shape brings the cells to the attention of the immune system, which destroys them along with the parasites they harbor. By contrast, *Plasmodium*-infected red blood cells of individuals homozygous for the *HbA* allele do not sickle, so the parasite may remain hidden from the immune system.

In areas where malaria is common, the persistence of the *HbS* allele is a matter of relative evils. Malaria and sickle-cell anemia are both potentially deadly. Heterozygous individuals may not be completely healthy, but they do have a better chance of surviving malaria than people homozygous for the normal allele (*HbA/HbA*). With or without malaria, people who have both alleles (*HbA/HbS*) are more likely to live long enough to reproduce than individuals homozygous for the sickle allele (*HbS/HbS*). The result is that nearly one-third of people living in the most malaria-ridden regions of the world carry the *HbS* allele.

balanced polymorphism Maintenance of two or more alleles of a gene at high frequency in a population.
frequency-dependent selection Natural selection in which a trait's adaptive value depends on its frequency in a population.
sexual dimorphism Difference in appearance between males and females of a species.
sexual selection Mode of natural selection in which some individuals outreproduce others of a population because they are better at securing mates.

TAKE-HOME MESSAGE 17.7
How does natural selection maintain diversity?

✔ With sexual selection, adaptive forms of a trait are those that give an individual an advantage in securing mates.

✔ Sexual selection can reinforce phenotypic differences between males and females, and sometimes it results in exaggerated traits.

✔ Balanced polymorphism can be an outcome of frequency-dependent selection, or of environmental pressures that favor heterozygous individuals.

✔ Especially in small populations, random changes in allele frequencies can lead to a loss of genetic diversity.

✔ Interbreeding among populations can change or stabilize allele frequencies, as can individuals (along with their alleles) moving from one population to another.

Genetic Drift

Genetic drift is random change in an allele's frequency over time, brought about by chance alone. We explain genetic drift in terms of probability (the chance that some event will occur, Section 1.8). Sample size is important in probability. Each time you flip a coin, there is a 50 percent chance it will land heads up. With 10 flips, the proportion of times heads actually land up may be very far from 50 percent. With 1,000 flips, that proportion is more likely to be near 50 per-

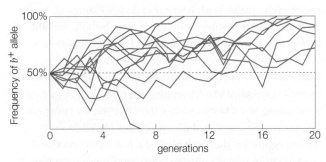

A The size of these populations was maintained at 10 breeding individuals. Allele b^+ was lost in one population (one graph line ends at 0).

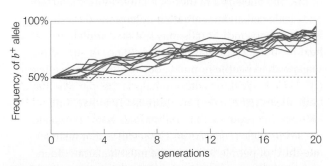

B The size of these populations was maintained at 100 individuals. Drift in these populations was less than in the small populations in **A**.

FIGURE 17.14 ▶Animated Genetic drift experiment in flour beetles (*Tribolium castaneum*), shown left on a flake of cereal.

Beetles heterozygous for alleles b^+ and b were maintained in populations of (**A**) 10 individuals or (**B**) 100 individuals for 20 generations. Graph lines in **B** are smoother than in **A**, indicating that drift was greatest in the sets of 10 beetles and least in the sets of 100.

Notice that the average frequency of allele b^+ rose at the same rate in both groups, an indication that natural selection was at work too: Allele b^+ was weakly favored. **FIGURE IT OUT** In how many populations did allele b^+ become fixed?   Answer: Six

cent. The same rule holds for populations: the larger the population, the smaller the impact of random changes in allele frequencies. Imagine two populations, one with 10 individuals, the other with 100. If allele X occurs in both populations at a 10 percent frequency, then only one person carries the allele in the small population. If that individual dies without reproducing, then the population's gene pool will lose allele X. However, ten individuals in the large population carry the allele. All ten would have to die without reproducing for the allele to be lost. Thus, the chance that the small population will lose allele X is greater than that for the large population. This is a general effect: The loss of genetic diversity is possible in all populations, but it is more likely to occur in small ones (**FIGURE 17.14**). When all individuals of a population are homozygous for an allele, we say that the allele is **fixed**. The frequency of a fixed allele will not change unless a new mutation occurs, or an individual bearing another allele enters the population.

Bottlenecks and the Founder Effect

A drastic reduction in population size, which is called a **bottleneck**, can greatly reduce genetic diversity. For example, northern elephant seals (shown in **FIGURE 17.12A**) underwent a bottleneck during the late 1890s, when hunting reduced their population size to about twenty individuals. Hunting restrictions have since allowed the population to recover, but genetic diversity among its members has been greatly reduced. The bottleneck and subsequent genetic drift eliminated many alleles that had previously been present in the population.

A loss of genetic diversity can also occur when a small group of individuals establishes a new population. If the founding group is not representative of the original population in terms of allele frequencies, then the new population will not be representative of it either. This outcome is called the **founder effect** (**FIGURE 17.15A**). Consider that all three *ABO* alleles for blood type (Section 13.5) are common in most human populations. Native Americans are an exception, with the majority of individuals being homozygous for the *O* allele. Native Americans are descendants of early humans who migrated from Asia

bottleneck Reduction in population size so severe that it reduces genetic diversity.
fixed Refers to an allele for which all members of a population are homozygous.
founder effect After a small group of individuals founds a new population, allele frequencies in the new population differ from those in the original population.
gene flow The movement of alleles into and out of a population.
genetic drift Change in allele frequency due to chance alone.
inbreeding Mating among close relatives.

CREDITS: (14A, B) Adapted from S. S. Rich, A. E. Bell, and S. P. Wilson, "Genetic drift in small populations of Tribolium," *Evolution* 33:579–584, Fig. 1, p. 580, © 1979 by John Wiley and Sons; left photo, Peggy Greb/USDA.

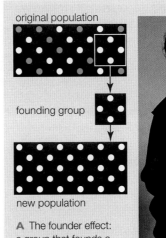

original population

founding group

new population

A The founder effect: a group that founds a new population is not representative of the original population, so allele frequencies differ between the new and the old populations.

B A high frequency of an allele that causes Ellis–van Creveld syndrome among the Lancaster Amish began with the founder effect.

FIGURE 17.15 The founder effect and one outcome.

between 14,000 and 21,000 years ago, across a narrow land bridge that once connected Siberia and Alaska. Analysis of DNA from ancient skeletal remains reveals that most early Americans were also homozygous for the *O* allele. Modern Siberian populations have all three alleles. Thus, the first humans in the Americas were probably members of a small group that had reduced genetic diversity compared with the general population.

Founding populations are often necessarily inbred. **Inbreeding** is mating between close relatives. Closely related individuals tend to share more alleles than nonrelatives do, so inbred populations often have unusually high numbers of individuals homozygous for recessive alleles, some of which are harmful. This outcome is minimized in human populations that discourage or forbid incest (mating between parents and children or between siblings).

The Old Order Amish in Lancaster County, Pennsylvania, offer an example of the effects of inbreeding. Amish people marry only within their community. Intermarriage with other groups is not permitted, and no "outsiders" are allowed to join the community. As a result, Amish communities are moderately inbred, and many of their individuals are homozygous for harmful recessive alleles. The Lancaster community has an unusually high frequency of a recessive allele that causes Ellis–van Creveld syndrome, a genetic disorder characterized by dwarfism,

polydactyly, and heart defects, among other symptoms. This allele has been traced to a man and his wife, two of a group of 400 Amish people who immigrated to the United States in the mid-1700s. As a result of the founder effect and inbreeding since then, about 1 of 8 people in the Lancaster community is now heterozygous for the allele, and 1 in 200 is homozygous for it (**FIGURE 17.15B**).

Gene Flow

Individuals of natural populations tend to mate or breed most frequently with other members of their own population. However, not all populations of a species are completely isolated from one another, and nearby populations may occasionally interbreed. Also, individuals sometimes leave one population and join another. **Gene flow**, the movement of alleles between populations, occurs in both cases. Gene flow can change or stabilize allele frequencies, thus countering the evolutionary effects of mutation, natural selection, and genetic drift.

Gene flow is typical among populations of animals, but it also occurs in less mobile organisms. Consider the acorns that jays disperse when they gather nuts for the winter (left). Every fall, these birds visit acorn-bearing oak trees repeatedly, then bury the acorns in the soil of territories as much as a mile away. The jays transfer acorns (and the alleles carried by these seeds) among populations of oak trees that may otherwise be genetically isolated. Gene flow also occurs when wind or an animal transfers pollen from one plant to another, often over great distances (more about this in Chapter 29). Many opponents of genetic engineering cite gene flow from transgenic crop plants into wild populations via pollen transfer. For example, engineered genes that confer resistance to herbicides and Bt (Section 15.7) are now commonly found in weeds and unmodified crop plants. Long-term effects of this gene flow are currently unknown.

TAKE-HOME MESSAGE 17.8

Other than natural selection, what mechanisms affect allele frequencies?

✔ Genetic drift can reduce a population's genetic diversity. Its effect is greatest in small populations.

✔ A population's genetic diversity may be lowered after a bottleneck, or as a result of the founder effect.

✔ Gene flow tends to oppose the evolutionary effects of mutation, natural selection, and genetic drift in a population.

17.9 Reproductive Isolation

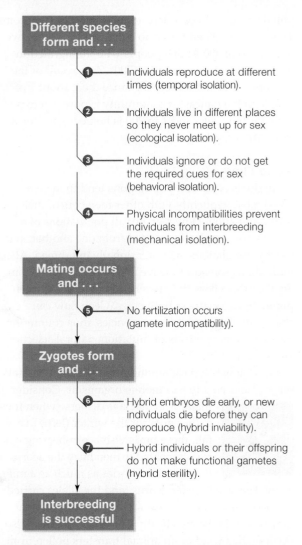

Different species form and . . .

1 — Individuals reproduce at different times (temporal isolation).

2 — Individuals live in different places so they never meet up for sex (ecological isolation).

3 — Individuals ignore or do not get the required cues for sex (behavioral isolation).

4 — Physical incompatibilities prevent individuals from interbreeding (mechanical isolation).

Mating occurs and . . .

5 — No fertilization occurs (gamete incompatibility).

Zygotes form and . . .

6 — Hybrid embryos die early, or new individuals die before they can reproduce (hybrid inviability).

7 — Hybrid individuals or their offspring do not make functional gametes (hybrid sterility).

Interbreeding is successful

FIGURE 17.16 ▶Animated How reproductive isolation prevents interbreeding.

FIGURE 17.17 Behavioral isolation. A male peacock spider (*Maratus volans*) approaches a female, signaling his intent to mate with her by raising and waving colorful flaps, and gesturing his legs in time with abdominal vibrations. If his species-specific courtship display fails to impress her, she will kill him.

✔ Speciation differs in its details, but reproductive isolation mechanisms are always part of the process.

When two populations do not interbreed, the number of genetic differences between them increases because mutation, natural selection, and genetic drift occur independently in each one. Over time, the populations may become so different that we consider them to be different species. Evolutionary processes in which new species arise are called **speciation**.

Evolution is a dynamic, extravagant, messy, and ongoing process that can be challenging for people who like categories. Speciation offers a perfect example, because it rarely occurs at a precise moment in time: Individuals often continue to interbreed even as populations are diverging, and populations that have already diverged may come together and interbreed again.

Every time speciation happens, it happens in a unique way, and each species is a product of its own unique evolutionary history. However, there are recurring patterns. For example, reproductive isolation is always part of speciation. **Reproductive isolation**, the end of gene flow between populations, is part of the process by which sexually reproducing species attain and maintain their separate identities. Mechanisms that prevent successful interbreeding reinforce differences between diverging populations (**FIGURE 17.16**).

❶ Some closely related species cannot interbreed because the timing of their reproduction differs (an effect called temporal isolation). Consider the periodi- cal cicada (left). Larvae of these insects feed on roots as they mature underground, then the adults emerge to reproduce. Three cicada species reproduce every 17 years. Each has a sibling species with nearly identical form and behavior, except that the siblings emerge on a 13-year cycle instead of a 17-year cycle. Sibling species have the potential to interbreed, but they can only get together once every 221 years!

❷ Adaptation to different microenvironments may prevent closely related species from interbreeding (ecological isolation). For example, two species of manzanita, a plant native to the Sierra Nevada mountain range, rarely hybridize. One species that lives on dry, rocky hillsides is better adapted for conserving water. The other, less drought-adapted species lives on lower slopes where water stress is not as intense. The physical separation makes cross-pollination unlikely.

❸ Differences in behavior can prevent mating between related animal species (behavioral isolation). For example, males and females of many animal species engage in courtship displays before sex (**FIGURE 17.17**). In a typical pattern, the female recognizes the sounds and movements of a male of her species as an overture to sex; females of different species do not.

❹ The size or shape of an individual's reproductive parts may prevent it from mating with members of closely related species (mechanical isolation). For example, plants called black sage and white sage grow in the same areas, but hybrids rarely form because the flowers of these two related species have become specialized for different pollinator species (**FIGURE 17.18**).

❺ Even if gametes of different species do meet up, they often have molecular incompatibilities that prevent a zygote from forming. For example, the molecular signals that trigger pollen germination in flowering plants are species-specific (we return to pollen germination and other aspects of flowering plant reproduction in Section 29.4). Gamete incompatibility may be the primary speciation route among animals that release their eggs and free-swimming sperm into water.

❻ Genetic changes are the basis of divergences in form, function, and behavior. Even chromosomes of species that diverged relatively recently may be different enough that a hybrid zygote inherits extra or missing genes, or genes with incompatible products—outcomes that typically disrupt development. Hybrids that do survive embryonic development often have reduced fitness. For example, hybrid offspring of lions and tigers have more health problems and a shorter life expectancy than individuals of either parent species.

❼ Some interspecies crosses produce robust but sterile hybrid offspring. For example, mating between a female horse (64 chromosomes) and a male donkey (62 chromosomes) produces a mule. Mules are healthy, but their 63 chromosomes cannot pair up evenly during meiosis, so this animal makes few viable gametes. If hybrids are fertile, their offspring usually have lower and lower fitness with each successive generation. Incompatible nuclear and mitochondrial DNA may be the cause (mitochondrial DNA is inherited from the mother only).

reproductive isolation The end of gene flow between populations.
speciation Evolutionary process in which new species arise.

TAKE-HOME MESSAGE 17.9
How do species attain and maintain separate identities?

✔ Speciation is an evolutionary process in which new species form. It varies in its details and duration, but reproductive isolation is always a part of speciation.

A Black sage is pollinated mainly by small insects.

B The flowers of black sage are too delicate to support larger insects. Big insects access the nectar of small sage flowers only by piercing from the outside, as this carpenter bee is doing. When they do so, they avoid touching the flower's reproductive parts.

C The reproductive parts (anthers and stigma) of white sage flowers are too far away from the petals to be brushed by honeybees, so honeybees are not efficient pollinators of this species. White sage is pollinated mainly by larger bees and hawkmoths, which brush the flower's stigma and anthers as they pry apart the petals to access nectar.

FIGURE 17.18 Mechanical isolation in sage.

CREDITS: (18A) Courtesy of Dr. James French; (18B) Courtesy of © Ron Brinkmann, www.flickr.com/photos/ronbrinkmann; (18C) © David Goodin.

✔ In allopatric speciation, a physical barrier arises and ends gene flow between populations.

Genetic changes that lead to a new species can begin with physical separation between populations. With **allopatric speciation**, a physical barrier arises and separates two populations, ending gene flow between them (*allo*– means different; *patria*, fatherland). Then, reproductive isolating mechanisms evolve that prevent interbreeding even if the diverging populations meet again.

Gene flow between populations separated by distance is often inconsistent. Whether a geographic barrier can completely block that gene flow depends on how the species travels (such as by swimming, walking, or flying), and how it reproduces (for example, by internal fertilization or by pollen dispersal).

A geographic barrier can arise in an instant, or over an eon. The Great Wall of China is an example of a barrier that arose abruptly. As it was being built, the wall interrupted gene flow among nearby populations of insect-pollinated plants; DNA sequence comparisons show that trees, shrubs, and herbs on either side of the wall are diverging genetically. Geographic isolation usually occurs much more slowly. For example, it took millions of years of tectonic plate movements (Section 16.7) to bring the two continents of North and South America close enough to collide. The land bridge where the two continents now connect is called the Isthmus of Panama. When this isthmus formed about

4 million years ago, it cut off the flow of water—and gene flow among populations of aquatic organisms—as it separated one large ocean into what are now the Pacific and Atlantic Oceans (**FIGURE 17.19**).

Speciation in Archipelagos

New species rarely form on island chains such as the Florida Keys that are in close proximity to a mainland. Being close to a mainland means gene flow is essentially unimpeded between island and mainland populations. By contrast, allopatric speciation is common on archipelagos (island chains) such as the Hawaiian and Galápagos Islands. These islands are so geographically isolated that, for most species, no gene flow occurs between island and mainland populations.

The Hawaiian archipelago includes 19 islands and more than 100 atolls stretching 1,500 miles in the Pacific Ocean. These islands are the product of hot spots on the ocean floor (Section 16.7). Because they were the tops of volcanoes, we can assume that their fiery surfaces were initially barren and inhospitable to life. Later, winds and currents carried individuals of mainland species to them. The individuals reproduced, and their descendants established populations. The lack of gene flow with mainland populations allowed the island populations to diverge. Today, thousands of species are unique to this island chain.

Consider Hawaiian honeycreepers, birds that are descendants of Asian finches that arrived on the

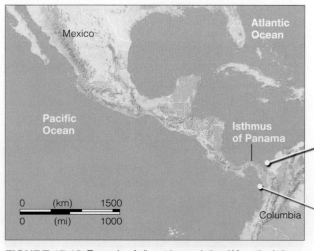

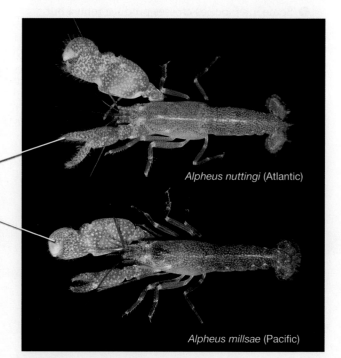

Alpheus nuttingi (Atlantic)

Alpheus millsae (Pacific)

FIGURE 17.19 Example of allopatric speciation. When the Isthmus of Panama formed, it cut off gene flow among ocean-dwelling populations of snapping shrimp. Today, shrimp on opposite sides of the isthmus might be able to interbreed were it not for behavioral isolation: Instead of mating when they are brought together, they snap their claws at one another aggressively. The photos show two of the many closely related species that live on opposite sides of the isthmus.

Akepa (*Loxops coccineus*) Akekee (*Loxops caeruleirostris*) Nihoa finch (*Telespiza ultima*) Palila (*Loxioides bailleui*)

Iiwi (*Vestiaria coccinea*) Akohekohe (*Palmeria dolei*) Apapane (*Himatione sanguinea*) Akiapolaau (*Hemignathus munroi*)

Maui parrotbill (*Pseudonestor xanthophrys*) Maui Alauahio (*Paroreomyza montana*) Kauai Amakihi (*Hemignathus kauaiensis*) Hawaii Amakihi (*Hemignathus virens*)

FIGURE 17.20 ▶Animated
Example of allopatric speciation on an archipelago. The ancestor of all Hawaiian honeycreepers was a rosefinch species (*Carpodacus*, left) from southern Asia. About 5.8 million years ago, a population of these finches somehow managed to fly thousands of miles across the open ocean to the Hawaiian archipelago (right). The expanse of ocean prevented gene flow between mainland populations and the island colonizers, which subsequently diverged into many honeycreeper species.

Hawaiian archipelago

islands at least 5.8 million years ago (**FIGURE 17.20**). A buffet of fruits, seeds, nectars, tasty insects, and the near absence of competitors and predators allowed the finch's descendants to thrive. In the absence of gene flow, the island finch population diverged from the ancestral mainland species. Further divergences occurred as new islands arose; habitats on the landmasses of the archipelago vary dramatically—from lava beds, rain forests, and grasslands to dry wood-lands and snow-capped peaks. Selection pressures differ within and between these habitats. The cumulative result of all these divergences is a spectacular array of honeycreeper species.

TAKE-HOME MESSAGE 17.10
What is allopatric speciation?

✔ A physical barrier that intervenes between populations of a species prevents gene flow among them. When gene flow ends, genetic divergences give rise to new species. This pattern is called allopatric speciation.

allopatric speciation Speciation pattern in which a physical barrier arises and ends gene flow between populations.

✔ Populations sometimes speciate even without a physical barrier that bars gene flow between them.

Sympatric Speciation

In **sympatric speciation**, populations inhabiting the same geographic region speciate in the absence of a physical barrier between them (*sym*– means together). Sympatric speciation can occur in a single generation when the chromosome number multiplies. Polyploidy (having three or more sets of chromosomes, Section 14.6) typically arises when an abnormal nuclear division during meiosis or mitosis doubles the chromosome number. For example, if the nucleus of a somatic cell in a flowering plant fails to divide during mitosis, the resulting cell—which is polyploid—may proliferate and give rise to shoots and flowers. If the flowers can self-fertilize, a new polyploid species may be the result. Common bread wheat originated after related species hybridized, and then the chromosome number of the hybrid offspring doubled (**FIGURE 17.21**).

Sympatric speciation can also occur with no change in chromosome number. The mechanically isolated sage plants you learned about in Section 17.9 speciated with no physical barrier to gene flow. As another example, more than 500 species of cichlid fishes arose by sympatric speciation in the shallow waters of Lake Victoria. This large freshwater lake sits isolated from river inflow on an elevated plain in Africa's Great Rift Valley. Since Lake Victoria formed about 400,000 years ago, it has dried up three times. DNA sequence com-

FIGURE 17.22 Red fish, blue fish: Males of four closely related species of cichlid native to Lake Victoria, Africa. Hundreds of cichlid species arose by sympatric speciation in this lake. Mutations that affect female cichlids' perception of the color of ambient light in deeper or shallower regions of the lake also affect their choice of mates. Female cichlids prefer to mate with brightly colored males of their own species.

FIGURE IT OUT Which form of natural selection is driving sympatric speciation in these cichlids?　　Answer: Sexual selection

parisons indicate that almost all of the cichlid species in this lake arose since the last dry spell, which was 12,400 years ago. How could hundreds of species arise so quickly? In this case, the answer begins with differences in the color of ambient light in different parts of the lake. The light in the lake's shallower, clear water is mainly blue; light that penetrates the deeper, muddier water is mainly red. The cichlid species vary in color

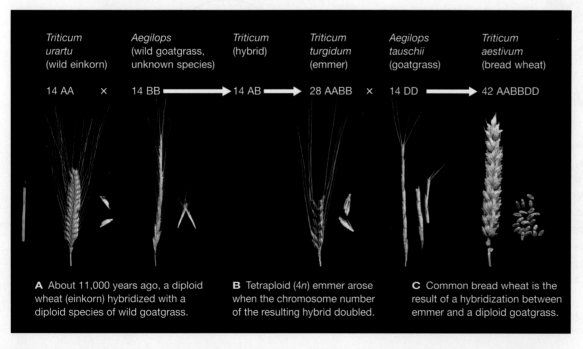

Triticum urartu (wild einkorn)	*Aegilops* (wild goatgrass, unknown species)	*Triticum* (hybrid)	*Triticum turgidum* (emmer)	*Aegilops tauschii* (goatgrass)	*Triticum aestivum* (bread wheat)
14 AA ×	14 BB →	14 AB →	28 AABB ×	14 DD →	42 AABBDD

A About 11,000 years ago, a diploid wheat (einkorn) hybridized with a diploid species of wild goatgrass.

B Tetraploid (4*n*) emmer arose when the chromosome number of the resulting hybrid doubled.

C Common bread wheat is the result of a hybridization between emmer and a diploid goatgrass.

FIGURE 17.21
▶**Animated**
Sympatric speciation in wheat. The wheat genome, which consists of seven chromosomes, occurs in slightly different forms called A, B, C, D, and so on. Many wheat species are polyploid, carrying more than two copies of the genome. For example, modern bread wheat (*Triticum aestivum*) is hexaploid, with six copies of the wheat genome: two each of genomes A, B, and D (or 42 AABBDD).

CREDITS: (21) Photos by © J. Honegger, courtesy of S. Stamp, E. Merz, www.sortengarten/ethz.ch; (22) Kevin Bauman, www.african-cichlid.com.

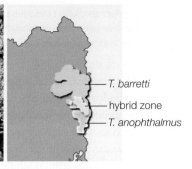

T. barretti

hybrid zone

T. anophthalmus

A Giant velvet walking worm, *Tasmanipatus barretti*.

B Blind velvet walking worm, *T. anophthalmus*.

C The habitats of the worms overlap in a hybrid zone on the island of Tasmania.

FIGURE 17.23 Example of parapatric speciation: velvet walking worms in Tasmania.

(**FIGURE 17.22**). Outside of captivity, female cichlids rarely mate with males of other species. Given a choice, they prefer to mate with brightly colored males of their own species. Their preference has a genetic basis, in alleles that encode light-sensitive pigments of the retina (part of the eye). Retinal pigments made by species that live mainly in shallow areas of the lake are more sensitive to blue light. The males of these species are also the bluest. Retinal pigments made by species that prefer deeper areas of the lake are more sensitive to red light. Males of these species are redder. In other words, the colors that a female cichlid sees best are the same colors displayed by males of her species. Thus, mutations that affect color perception are likely to affect a female's choice of mates. Such mutations are probably the way sympatric speciation occurs in these fishes.

Sympatric speciation has also occurred in greenish warblers of central Asia (*Phylloscopus trochiloides*). A chain of populations of this bird encircles the Tibetan 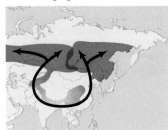 plateau (left). Adjacent populations of greenish warblers interbreed easily, except for the two populations at the ends of the chain. These two populations overlap in northern Siberia, but their individuals do not interbreed because they do not recognize one another's songs (an example of behavioral isolation). Small genetic differences between adjacent populations have added up to major differences between the two populations at the ends

of the chain. Greenish warbler populations that make up the chain are collectively called a ring species. Ring species present one of those paradoxes for people who like neat categories: Gene flow occurs continuously all around the chain, but the two populations at the ends of the chain are clearly different species. Where should we draw the line that divides those two species?

Parapatric Speciation

With **parapatric speciation**, adjacent populations speciate despite being in contact across a common border. Divergences spurred by local selection pressures are reinforced because hybrids that form in the contact zone are less fit than individuals on either side of it.

Consider velvet walking worms, which resemble caterpillars but may be more related to spiders: They are predatory, and shoot streams of glue from their head to entangle insect prey. Two rare species of velvet walking worm are native to the island of Tasmania (**FIGURE 17.23**). The giant velvet walking worm and the blind velvet walking worm can interbreed, but they only do so in a tiny area where their habitats overlap. Hybrid offspring are sterile, which may be the main reason the two species are maintaining separate identities in the absence of a physical barrier between their adjacent populations.

---

parapatric speciation Populations inhabiting different areas speciate while in contact along a common border.
sympatric speciation Divergence within a population leads to speciation; occurs in the absence of a physical barrier to gene flow.

TAKE-HOME MESSAGE 17.11

Does speciation occur in the absence of a physical barrier to gene flow?

✔ With sympatric speciation, divergence within a population leads to new species that inhabit the same geographical area, with no physical barrier to gene flow.

✔ With parapatric speciation, populations maintaining contact along a common border evolve into distinct species.

✔ Macroevolution includes patterns of change such as one species giving rise to many, the origin of major groups, and major extinction events.

Microevolution is change in allele frequencies within a single species or population. **Macroevolution** is our name for evolutionary patterns on a larger scale: trends such as land plants evolving from green algae, the dinosaurs disappearing in a mass extinction, a burst of divergences from a single species, and so on.

The simplest macroevolutionary pattern is **stasis**, in which little change occurs over a very long period of time. Consider coelacanths, an order of ancient lobe-finned fish that had been assumed extinct for at least 70 million years until a fisherman caught one in 1938. In form and other aspects, modern coelacanth species are similar to fossil specimens hundreds of millions of years old (**FIGURE 17.24**).

Major evolutionary novelties often stem from the adaptation of an existing structure for a completely new purpose. This macroevolutionary pattern is called **exaptation**. For example, the feathers that allow modern birds to fly are derived from feathers that first evolved in some dinosaurs. Those dinosaurs could not have used their feathers for flight, but they probably did use them for insulation. Thus, we say that flight feathers in birds evolved by exaptation from insulating feathers in dinosaurs.

By current estimates, more than 99 percent of all species that ever lived are now **extinct**, which means they no longer have living members. In addition to continuing small-scale extinctions, the fossil record indicates that there have been more than twenty mass extinctions, which are simultaneous losses of many lineages. These include five catastrophic events in which the majority of species on Earth disappeared (Section 16.8).

With **adaptive radiation**, one lineage rapidly diversifies into several new species. An adaptive radiation typically occurs after a population colonizes a new environment that has a variety of different habitats and few or no competitors. Speciation occurs along with adaptation to the different habitats. The Hawaiian honeycreepers arose this way, as did the Lake Victoria cichlids. Adaptive radiation may also occur after a key innovation evolves. A **key innovation** is a new trait that allows its bearer to exploit a habitat more efficiently or in a novel way. The evolution of lungs offers an example, because lungs were a key innovation that opened the way for an adaptive radiation of vertebrates on land. A geologic or climatic event that eliminates some species from a habitat can spur adaptive radiation; species that survive the event then have access to resources from which they had previously been excluded. This is the way mammals were able to undergo an adaptive radiation after the dinosaurs disappeared.

The process by which close ecological interactions between two species cause them to evolve jointly is called **coevolution**. One species acts as an agent of selection on the other, and each adapts to changes in the other. Over evolutionary time, the two species may become so interdependent that they can no longer survive without one another.

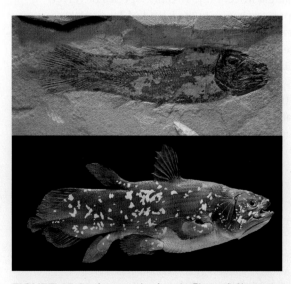

Notochord
This tough, elastic tube, which is partially hollow and filled with fluid, is ancestral to the spinal cord.

Lobed fins
These fleshy fins retain a few of the ancestral bones that gave rise to legs and arms in other lineages.

Long gestation
Coelacanths give birth to litters of up to 26 fully developed "pups" after gestation of more than a year.

Rostral organ
A sensory organ that perceives electrical impulses in water, it probably helps the fish locate prey in dark ocean depths.

FIGURE 17.24 An example of stasis. Photos (left) compare a 320-million-year-old coelacanth fossil found in Montana with a live coelacanth. The diagram (right) shows a few of the coelacanth's unusual ancestral features that have been lost in almost all other fish lineages over evolutionary time.

Antibiotic-resistant bacteria bred inside people or treated animals end up in the environment, where they can easily spread to other individuals. These bacteria have plagued hospitals for years, and now we find them everywhere. They are common in day-care centers, schools, gyms, prisons, and other places where people are in close contact. We also find them in our pets, and in wildlife such as crows, rabbits, mongooses, foxes, sharks, rodents, reptiles, birds, frogs, whales, chimpanzees, penguins, and insects such as moths and houseflies. They have even been found in beach sand, coastal waters, and Antarctic seawater.

Relationships between coevolved species can be quite intricate. Consider the large blue butterfly (*Maculinea arion*), a parasite of ants. After hatching, the butterfly larvae (caterpillars) feed on wild thyme flowers and then drop to the ground. An ant that finds a caterpillar strokes it, which makes the caterpillar exude honey. The ant eats the honey and continues to stroke the caterpillar, which secretes more honey. This interaction continues for about an hour, until the caterpillar suddenly hunches itself up (**FIGURE 17.25**). The ant then picks up the caterpillar and carries it back to its nest, where, in most cases, other ants kill it—except if the ants are of the species *Myrmica sabuleti*. The caterpillar secretes the same chemicals as *Myrmica sabuleti* larvae, and makes the same sounds as their queen—behaviors that deceive the ants into adopting the caterpillar and treating it better than their own larvae. The adopted caterpillar feeds on ant larvae for about 10 months, then undergoes metamorphosis, changing into a butterfly that emerges from the ground to mate. Eggs are deposited on wild thyme near another *M. sabuleti* nest, and the cycle starts anew. This relationship between ant and butterfly is typical of coevolved relationships in that it is extremely specific. Any increase in the ants' ability to identify a caterpillar in their nest selects for caterpillars that better deceive the ants, which in turn select for ants that can better identify the caterpillars. Each species exerts directional selection on the other.

FIGURE 17.25 A *Myrmica sabuleti* ant preparing to carry a hunched-up *Maculinea arion* caterpillar back to its nest. If adopted by the ant colony, the caterpillar will eat ant larvae until it matures.

adaptive radiation A burst of genetic divergences from a lineage gives rise to many new species.
coevolution The joint evolution of two closely interacting species; each species is a selective agent for traits of the other.
exaptation Evolutionary adaptation of an existing structure for a completely new purpose.
extinct Refers to a species that no longer has living members.
key innovation An evolutionary adaptation that gives its bearer the opportunity to exploit a particular environment much more efficiently or in a new way.
macroevolution Large-scale evolutionary patterns and trends.
stasis Evolutionary pattern in which little or no change occurs over long spans of time.

Evolutionary Theory

Biologists do not doubt that macroevolution occurs, but many disagree about how it occurs. However we choose to categorize evolutionary processes, the very same genetic change may be at the root of all evolution—fast or slow, large-scale or small-scale. Dramatic jumps in morphology, if they are not artifacts of gaps in the fossil record, may be the result of mutations in homeotic or other regulatory genes. Macroevolution may include more processes than microevolution, or it may not. It may be an accumulation of many microevolutionary events, or it may be an entirely different process. Evolutionary biologists may disagree about these and other hypotheses, but all of them are trying to explain the same thing: how all species are related by descent from common ancestors.

TAKE-HOME MESSAGE 17.12
What is macroevolution?

✔ Macroevolution comprises large-scale patterns of evolutionary change such as adaptive radiation, the origin of major groups, and mass extinctions.

summary

Section 17.1 Our overuse of antibiotics exerts directional selection favoring resistant bacterial populations, which are now common in the environment. We are running out of effective antibiotics to use as human drugs.

Sections 17.2, 17.3 All alleles of all genes in a **population** constitute a **gene pool**. Mutations may be **neutral**, **lethal**, or adaptive. **Microevolution** is change in **allele frequency** of a population. Deviations from **genetic equilibrium** indicate that a population is evolving.

Sections 17.4–17.6 In **directional selection**, a phenotype at one end of a range of variation is adaptive. An intermediate form of a trait is adaptive in **stabilizing selection**; extreme forms are adaptive in **disruptive selection**.

Section 17.7 **Sexual dimorphism** is a potential outcome of **sexual selection**, a mode of natural selection in which adaptive traits are those that make their bearers better at securing mates. **Frequency-dependent selection** or any other mode of natural selection can give rise to a **balanced polymorphism**.

Section 17.8 **Genetic drift**, which is most pronounced in small or **inbreeding** populations, can cause alleles to become **fixed**. The **founder effect** may occur after an evolutionary **bottleneck**. **Gene flow** can counter the effects of mutation, natural selection, and genetic drift.

Section 17.9 The details of **speciation** differ every time it occurs, but **reproductive isolation**, the end of gene flow between populations, is always a part of the process (**TABLE 17.2**). The moment at which two populations become separate species is often impossible to pinpoint.

Table 17.2 Comparison of Speciation Models

	Allopatric	Parapatric	Sympatric
Original population(s)	●	●●	●
Initiating event:	physical barrier arises	selection pressures differ	genetic change
	◐	◐	●
Reproductive isolation occurs	◐	◐	●
New species arises:	in isolation	in contact along common border	within existing population
	●●	●●	●

Section 17.10 In **allopatric speciation**, a geographic barrier arises and interrupts gene flow between populations. After gene flow ends, genetic divergences occur independently in each population, and this can result in separate species.

Section 17.11 Speciation can occur in the absence of a barrier to gene flow. **Sympatric speciation** occurs by divergence within a population. Polyploid species of many plants (and a few animals) have originated this way. With **parapatric speciation**, populations in physical contact along a common border speciate.

Section 17.12 **Macroevolution** refers to large-scale patterns of evolution. With **stasis**, little or no change occurs over long spans of time. In **exaptation**, a lineage uses a structure for a different purpose than its ancestor. A **key innovation** can result in an **adaptive radiation**. **Coevolution** occurs when two species act as agents of selection upon one another. A lineage with no more living members is **extinct**.

self-quiz

Answers in Appendix VII

1. _____ is the original source of new alleles.
   - a. Mutation
   - b. Natural selection
   - c. Genetic drift
   - d. Gene flow
   - e. All are original sources of new alleles

2. Which is required for evolution to occur in a population?
   - a. random mating
   - b. selection pressure
   - c. gene flow
   - d. none of the above

3. Match the modes of natural selection with their best descriptions.
   - ___ stabilizing
   - ___ disruptive
   - a. eliminates extreme forms of a trait
   - b. eliminates midrange form of a trait

4. Sexual selection frequently influences aspects of body form and can lead to _____ .
   - a. a sexual dimorphism
   - b. male aggression
   - c. exaggerated traits
   - d. all of the above

5. The persistence of sickle-cell anemia in a population with a high incidence of malaria is a case of _____ .
   - a. bottlenecking
   - b. inbreeding
   - c. balanced polymorphism
   - d. the founder effect
   - e. frequency-dependent selection

6. _____ tends to keep populations of a species similar to one another.
   - a. Genetic drift
   - b. Gene flow
   - c. Mutation
   - d. Natural selection

7. The theory of natural selection does not explain _____ .
   - a. genetic drift
   - b. the founder effect
   - c. gene flow
   - d. how mutations arise
   - e. inheritance
   - f. any of the above

Resistance to Rodenticides in Wild Rat Populations Beginning in 1990, rat infestations in northwestern Germany started to intensify despite continuing use of rat poisons. In 2000, Michael H. Kohn and his colleagues analyzed the genetics of wild rat populations around Münster. For part of their research, they trapped wild rats in five towns, and tested those rats for resistance to warfarin and the more recently developed poison bromadiolone. The results are shown in **FIGURE 17.26**.

1. In which of the five towns were most of the rats susceptible to warfarin?

2. Which town had the highest percentage of poison-resistant wild rats?

3. What percentage of rats in Olfen were resistant to warfarin?

4. In which town do you think the application of bromadiolone was most intensive?

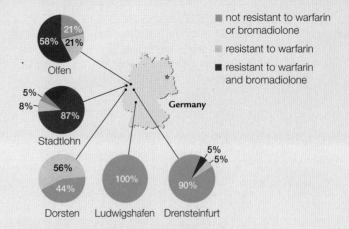

- ■ not resistant to warfarin or bromadiolone
- ▨ resistant to warfarin
- ■ resistant to warfarin and bromadiolone

FIGURE 17.26 Resistance to rat poisons in wild populations of rats in Germany, 2000.

8. Which of the following is *not* part of how we define a species?
   - a. Its individuals appear different from other species.
   - b. It is reproductively isolated from other species.
   - c. Its populations can interbreed.
   - d. Fertile offspring are produced.

9. Sex in many birds is typically preceded by an elaborate courtship dance. If a male's movements are unrecognized by the female, she will not mate with him. This is an example of _____ .
   - a. reproductive isolation
   - b. behavioral isolation
   - c. sexual selection
   - d. all of the above

10. The difference between sympatric and parapatric speciation is _____ .
   - a. parapatric speciation occurs only in worms
   - b. sympatric speciation requires a barrier to gene flow
   - c. the extent of overlap in range
   - d. reproductive isolation does not occur

11. A fire devastates all trees in a wide swath of forest. Populations of a species of tree-dwelling frog on either side of the burned area diverge to become separate species. This is an example of _____ .
   - a. allopatric speciation
   - b. parapatric speciation
   - c. sympatric speciation
   - d. adaptive radiation

12. Match the evolution concepts.
   - __c__ gene flow
   - __d__ sexual selection
   - __e__ mutation
   - __b__ genetic drift
   - __a__ coevolution
   - __f__ adaptive radiation
   - a. can lead to interdependent species
   - b. changes in a population's allele frequencies due to chance alone
   - c. alleles enter or leave a population
   - d. adaptive traits make their bearers better at securing mates
   - e. original source of new alleles
   - f. burst of divergences from one lineage into many

13. Change in allele frequency of a population is called _____ .
   - a. macroevolution
   - b. adaptive radiation
   - c. inbreeding
   - d. microevolution

critical thinking

1. Species have been traditionally characterized as "primitive" and "advanced." For example, mosses were considered to be primitive, and flowering plants advanced; crocodiles were primitive and mammals were advanced. Why do most biologists of today think it is incorrect to refer to any modern species as primitive?

2. Rama the cama, a llama–camel hybrid, was born in 1997. The idea was to breed an animal that has the camel's strength and endurance, and the llama's gentle disposition. However, instead of being large, strong, and sweet, Rama is smaller than expected and has a camel's short temper. The breeders plan to mate him with Kamilah, a female cama. What potential problems with this mating should the breeders anticipate?

3. Two species of antelope, one from Africa, the other from Asia, are put into the same enclosure in a zoo. To the zookeeper's surprise, individuals of the different species begin to mate and produce healthy, hybrid baby antelopes. Explain why a biologist might not view these offspring as evidence that the two species of antelope are in fact one.

4. Some people think that many of our uniquely human traits arose by sexual selection. Over thousands of years, women attracted to charming, witty men perhaps prompted the development of human intellect beyond what was necessary for mere survival. Men attracted to women with juvenile features may have shifted the species as a whole to be less hairy and softer featured than any of our simian relatives. Can you think of a way to test these hypotheses?

LEARNING ROADMAP

This chapter adds the concept of evolution (Sections 16.3 and 16.4) to taxonomy (1.5). Before starting, you should review DNA sequences (8.3) and sequencing (15.4); the genetic code (9.4); master genes (10.3, 10.4); genomics (15.5), neutral mutations (17.2), genetic equilibrium (17.3); gene flow (17.8); and speciation (17.9).

PHYLOGENY

Evolutionary biologists can reconstruct the evolutionary history of a group of organisms by identifying shared, heritable traits that evolved in a common ancestor.

COMPARING BODY FORM

Similar body parts in different lineages may indicate descent from a shared ancestor, or they may have evolved independently in response to similar environmental pressures.

COMPARING BIOCHEMISTRY

Neutral changes tend to accumulate at a fairly constant rate in DNA. Molecular comparisons help us discover and confirm relationships among species and lineages.

COMPARING DEVELOPMENT

Patterns of development have a basis in master genes conserved over evolutionary time. Lineages with more recent common ancestry often develop in similar ways.

APPLICATIONS OF PHYLOGENY

Understanding a group's evolutionary history can help us protect endangered species. Applied to agents of infectious disease, it can also reveal large-scale patterns of transmission.

You will revisit cladistics throughout Unit IV as you learn about life's diversity. Classification of birds and reptiles is explained in Chapter 25. Chapter 44 returns to population ecology, the study of which is important for understanding human impacts on the biosphere (Chapter 48). Our efforts to mitigate that impact include conservation efforts aimed at maintaining biodiversity (Chapter 48) in ecosystems (Chapter 46). Animal development is detailed in Chapter 42.

18.1 Bye Bye Birdie

Some finch species migrate far outside of their normal range when food becomes scarce in a preferred overwintering spot, traveling in flocks of thousands or even tens of thousands of individuals. About 5.8 million years ago in southern Asia, one of these migratory flocks was caught up in the winds of a huge storm. The birds—rosefinches—were blown at least seven thousand miles (11,000 kilometers) across the open ocean to the islands of the Hawaiian archipelago. Enough individuals survived the journey to found a new population. The birds' arrival had been preceded by insects and plants, but no predators, and their descendants thrived. Isolation from mainland finch populations allowed the island colonizers to diverge in adaptive radiations that gave rise to the Hawaiian honeycreepers (Section 17.10).

The first Polynesians arrived on the islands sometime before 1000 A.D.; Europeans followed in 1778. Hawaii's rich ecosystem was hospitable to the newcomers and their domestic animals and crops. Escaped livestock ate and trampled rain forest plants that had provided the honeycreepers with food and shelter. Entire forests were cleared to grow imported crops, and plants that escaped cultivation began to crowd out native plants. Mosquitoes accidentally introduced in 1826 spread diseases such as avian malaria from imported chickens to native bird species. Stowaway rats ate their way through populations of native birds and their eggs. Mongooses deliberately imported to eat the rats preferred to eat birds and bird eggs.

The isolation that had allowed honeycreepers to arise by adaptive radiation also made them vulnerable to extinction. Divergence from the ancestral species had led to the loss of unnecessary traits such as defenses against mainland predators and diseases. Traits that had previously been adaptive—such as a long, curved beak matching the flower of a particular plant—became hindrances when habitats suddenly changed or disappeared. Thus, at least 43 Hawaiian honeycreeper species that had thrived on the islands before humans arrived were extinct by 1778. Conservation efforts began in the 1960s, but another 43 species have since disappeared.

Today, the few remaining Hawaiian honeycreepers are still pressured by established populations of nonnative species of plants and animals. Rising global temperatures are also allowing mosquitoes to invade high-altitude habitats that had previously been too cold for the insects, so honeycreeper species remaining in these habitats are now succumbing to mosquito-borne diseases. Of the 18 remaining honeycreeper species, only two are not in danger of extinction (**FIGURE 18.1**).

A A Palila (*Loxioides bailleui*) feeds on the seeds of the mamane plant, which are toxic to most other birds. The one remaining Palila population is declining because mamane plants are being trampled by cows and gnawed to death by goats and sheep. About 2,176 Palila remained in 2012.

B The unusual skewed bill of the Akekee (*Loxops caeruleirostris*) allows this bird to easily pry open buds that harbor insects. Avian malaria carried by mosquitoes to higher altitudes is decimating the last population of this species. In 2008, about 3,111 remained.

C This male Poouli (*Melamprosops phaeosoma*)—rare, old, and missing an eye—died in 2004 from avian malaria. There were two other Poouli alive at the time, but neither has been seen since then.

FIGURE 18.1 Three honeycreeper species: going, going, gone.

CREDITS: (opposite) John Steiner/Smithsonian Institution; (1A) © Eric VanderWerf/Pacific Rim Photos; (1B) © Courtesy of © Lucas Behnke; (1C) Bill Sparklin/Ashley Dayer.

✔ Evolutionary history can be reconstructed by studying shared, heritable traits.

Table 18.1	Examples of Characters		
	Bird	**Bat**	**Lion**
Warm-blooded	Y	Y	Y
Hair	N	Y	Y
Milk	N	Y	Y
Teeth	N	Y	Y
Wings	Y	Y	N
Feathers	Y	N	N

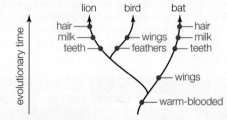

A If the bird and lion are most closely related, the derived traits would have evolved ten times in total.

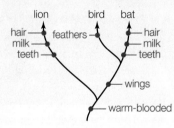

B If the bird and bat are most closely related, the derived traits would have evolved nine times in total.

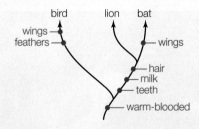

C If the lion and bat are most closely related, the derived traits would have evolved seven times in total.

FIGURE 18.2 An example of cladistics, using parsimony analysis with the characters listed in **TABLE 18.1**.

A, B, and **C** show the three possible evolutionary pathways that could connect birds, bats, and lions; red indicates the evolution of a derived trait. The pathway most likely to be correct (**C**) is the simplest—the one in which the derived traits would have had to evolve the fewest number of times in total.

Classifying life's tremendous diversity into a series of taxonomic ranks (Section 1.5) is a useful endeavor, in the same way that it is useful to organize a telephone book or contact list in alphabetical order. Today, biologists try to classify organisms according to evolutionary relationships among them. Thus, they focus on reconstructing **phylogeny**, the evolutionary history of a species or a group of species. Phylogeny is a kind of genealogy that follows a lineage's evolutionary relationships through time.

Humans were not around to witness the evolution of most species, but we can use evidence to understand events in the past (Section 16.1). Consider how each species bears traces of its own unique evolutionary history in its characters. A **character** is a quantifiable, heritable trait such as the number of segments in a backbone, the nucleotide sequence of ribosomal RNA, or the presence of wings (**TABLE 18.1**). Traditional classification schemes group organisms based on shared characters: Birds have feathers, cacti have spines, and so on. Such schemes do not necessarily reflect evolutionary history because species that are not closely related may appear very similar (Section 16.2). By contrast, evolutionary biology tries to fit each species into a bigger picture of evolution: Every living thing is related if you just go back far enough in time. Instead of grouping organisms by shared characters, evolutionary biologists try to pinpoint what makes the organisms share the characters in the first place: a common ancestor. They determine common ancestry by looking for derived traits. A **derived trait** is a character present in a group under consideration, but not in any of the group's ancestors.

A group whose members share one or more defining derived traits is called a **clade**. By definition, a clade is a **monophyletic group**: one that consists of an ancestor (in which a derived trait evolved) together with any and all of its descendants.

Each species is a clade. Many higher taxonomic rankings are also equivalent to clades—flowering plants, for example, are both a phylum and a clade—but some are not. For example, the traditional Linnaean class Reptilia ("reptiles") includes crocodiles, alligators, tuataras, snakes, lizards, turtles, and tortoises. While it is convenient to classify these animals together, they would not constitute a clade unless birds are also included, as you will see in Chapter 25.

It is the recent nature of a derived trait that defines a clade. Consider how alligators look a lot more like lizards than birds. In this case, the similarity in appearance does indicate shared ancestry, but it is a more distant relationship than alligators have with birds. Evolutionary biologists discovered that alligators and birds share a more recent common ancestor than alligators and lizards do. Derived traits—a gizzard and a four-chambered heart—evolved in the lineage that gave rise to alligators and birds, but not in the one that gave rise to lizards.

The remaining sections of this chapter explore some of the character comparisons that evolutionary biologists use to group organisms into clades. Remember that evolutionary history does not change because of events in the present: A species' ancestry remains the same no matter how it evolves. However, as with traditional taxonomy, we can make mistakes grouping organisms into clades if the information we have is incomplete. A clade is necessarily a hypothesis, and which organisms it includes may change when new discoveries are made. As with all hypotheses, the more data that support a cladistic grouping, the less likely it is to require revision.

Cladistics

In the big picture of evolution, all clades are interconnected; an evolutionary biologist's job is to figure out where the connections are. Making hypotheses about evolutionary relationships among clades is called **cladistics**. One way of doing this involves the logical rule of simplicity: When there are several possible ways that a group of clades can be connected, the simplest evolutionary pathway is probably the correct one. By comparing all of the possible connections among the clades, we can identify the simplest: the one in which the defining derived traits evolved the fewest number of times (**FIGURE 18.2**). The process of finding the simplest pathway is called parsimony analysis.

The result of a cladistic analysis is an **evolutionary tree**—a diagram of evolutionary connections—called a cladogram. A **cladogram** visually summarizes a

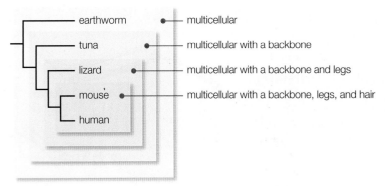

A Evolutionary connections among clades are represented as lines on a cladogram. Sister groups emerge from a node, which represents a common ancestor.

B A cladogram can be viewed as "sets within sets" of derived traits. Each set (an ancestor together with all of its descendants) is a clade.

FIGURE 18.3 ▶**Animated** An example of a cladogram.

hypothesis about how a group of clades are related (**FIGURE 18.3**). Data from an outgroup (a species not closely related to any member of the group under study) may be included in order to "root" the tree. Each line in a cladogram represents a lineage, which may branch into two lineages at a node.

The node represents a common ancestor of two lineages. Every branch of a cladogram is a clade; the two lineages that emerge from a node on a cladogram are called **sister groups**.

character Quantifiable, heritable characteristic or trait.
clade A group whose members share one or more defining derived traits.
cladistics Making hypotheses about evolutionary relationships among clades.
cladogram Evolutionary tree diagram that summarizes hypothesized relationships among a group of clades.
derived trait A novel trait present in a clade but not in any of the clade's ancestors.
evolutionary tree Diagram showing evolutionary connections.
monophyletic group An ancestor in which a derived trait evolved, together with all of its descendants.
phylogeny Evolutionary history of a species or group of species.
sister groups The two lineages that emerge from a node on a cladogram.

> ### TAKE-HOME MESSAGE 18.2
> #### Why do we study evolutionary history?
> ✔ Evolutionary biologists study phylogeny in order to understand how all species are connected by shared ancestry.
>
> ✔ A clade is a monophyletic group whose members share one or more derived traits. Cladistics is a method of making hypotheses about evolutionary relationships among clades.
>
> ✔ Cladograms and other evolutionary tree diagrams are based on our understanding of the evolutionary history of a group of organisms.

✔ Physical similarities are often evidence of shared ancestry, but sometimes a trait evolves independently in different lineages.

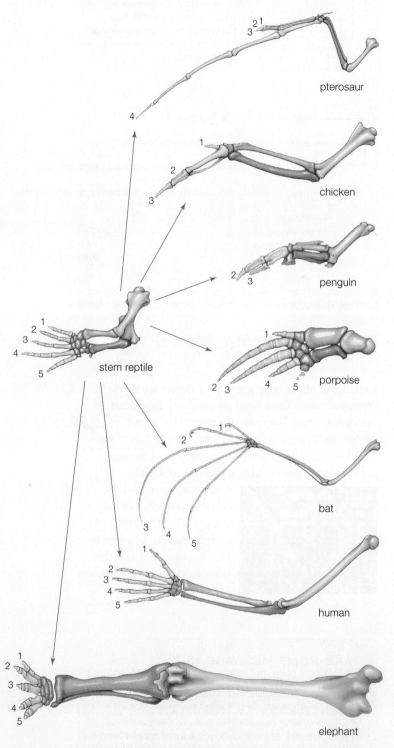

FIGURE 18.4 ▶Animated Morphological divergence among vertebrate fore-limbs, starting with the bones of an ancient stem reptile. The number and position of many skeletal elements were preserved when these diverse forms evolved; notice the bones of the forearms. Certain bones were lost over time in some of the lineages (compare the digits numbered 1 through 5). Drawings are not to scale.

To biologists, remember, evolution means change in a line of descent. How do they reconstruct evolutionary events that occurred in the ancient past? Evolutionary biologists are a bit like detectives, using clues to piece together a history that they did not witness in person. Fossils provide some clues. The body form and function of organisms that are alive today provide others.

Morphological Divergence

Body parts that appear similar in separate lineages because they evolved in a common ancestor are called **homologous structures** (*hom–* means "the same"). Homologous structures may be used for different purposes in different groups, but the very same genes direct their development.

A body part that outwardly appears very different in separate lineages may be homologous in underlying form. Vertebrate forelimbs, for instance, vary in size, shape, and function. However, they are alike in the structure and positioning of bony elements, and in internal patterns of nerves, blood vessels, and muscles.

You learned in Chapter 17 how populations that are not interbreeding diverge genetically; these divergences give rise to changes in body form. Change from the body form of a common ancestor is an evolutionary pattern called **morphological divergence**. Consider the limb bones of vertebrate animals. Fossil evidence suggests that many modern vertebrates are descended from a family of ancient "stem reptiles" that crouched low to the ground on five-toed limbs. Descendants of this ancestral group diversified over millions of years, and eventually gave rise to modern reptiles, birds, and mammals; a few lineages that had become adapted to walking on land even returned to aquatic living. As these lineages diversified, their five-toed limbs became adapted for many different purposes (**FIGURE 18.4**). Limbs became modified for flight in extinct reptiles called pterosaurs and in bats and most birds. In penguins and porpoises, they are now flippers useful for swimming. Forelimbs of humans are arms and hands with four fingers and an opposable thumb. Among elephants, the limbs are now strong and pillarlike, capable of supporting a great deal of weight. Limbs degenerated to nubs in pythons and boa constrictors, and they disappeared entirely in other snakes.

Morphological Convergence

Body parts that appear similar in different species are not always homologous; they sometimes evolve independently in lineages subject to the same environmental pressures. The independent evolution of similar body parts in different lineages is **morphological convergence**. Structures that are similar as a result of morphological convergence are called **analogous structures**. Analogous structures look alike but did not evolve in a shared ancestor; they evolved independently after the lineages diverged.

For example, bird, bat, and insect wings all perform the same function, which is flight. However, several clues tell us that the wing surfaces are not homologous. All of the wings are adapted to the same physical constraints that govern flight, but each is adapted in a different way. In the case of birds and bats, the limbs themselves are homologous, but the adaptations that make those limbs useful for flight differ. The surface of a bat wing is a thin, membranous extension of the animal's skin. By contrast, the surface of a bird wing is a sweep of feathers, which are specialized structures derived from skin. Insect wings differ even more. An insect wing forms as a saclike extension of the body wall. Except at forked veins, the sac flattens and fuses into a thin membrane. The sturdy, chitin-reinforced veins structurally support the wing. Unique adaptations for flight are evidence that wing surfaces of birds, bats, and insects are analogous structures that evolved after the ancestors of these modern groups diverged (**FIGURE 18.5**).

As another example of morphological convergence, consider the similar external structures of American cacti and African euphorbias (see **FIGURE 16.3**). These structures adapt the plants to similarly harsh desert environments where rain is scarce. Distinctive accordion-like pleats allow the plant body to swell with water when rain does come. Water stored in the plants' tissues allows them to survive long dry periods. As the stored water is used, the plant body shrinks, and the folded pleats provide it with some shade in an environment that typically has none. Despite these similarities, a closer look reveals many differences that indicate the two types of plants are not closely related. For example, cactus spines have a simple fibrous struc-

ture; they are modified leaves that arise from dimples on the plant's surface. Euphorbia spines project smoothly from the plant surface, and they are not modified leaves: In many species the spines are actually dried flower stalks (left).

FIGURE 18.5 Morphological convergence in animals. The surfaces of an insect wing, a bat wing, and a bird wing are analogous structures. The diagram shows how the evolution of wings (red dots) occurred independently in the three separate lineages.

---

analogous structures Similar body structures that evolved separately in different lineages.
homologous structures Body structures that are similar in different lineages because they evolved in a common ancestor.
morphological convergence Evolutionary pattern in which similar body parts evolve separately in different lineages.
morphological divergence Evolutionary pattern in which a body part of an ancestor changes in its descendants.

TAKE-HOME MESSAGE 18.3
What evidence does evolution leave in body form?

✔ In a pattern of morphological divergence, body parts are often modified differently in different lines of descent. Such parts are called homologous structures.

✔ In a pattern of morphological convergence, body parts that appear alike evolved independently in different lineages. Such parts are called analogous structures.

CREDITS: (in text) © James C. Gaither, www.flickr.com/people/jim-sf/; (5) top, © iStockphoto.com/DanCardiff; middle, © Taro Taylor, www.flickr.com/photos/tjt195; bottom, © Alberto J. Espiñeira Francés - Alesfra/Getty Images; art, © Cengage Learning.

18.4 Comparing Biochemistry

✔ Evolution leaves clues in biochemistry. In general, more closely related species share more biochemical similarities.

Over time, inevitable mutations change a genome's DNA sequence. Most of these mutations are neutral. Neutral mutations have no effect on an individual's survival or reproduction, so we can assume they accumulate at a constant rate. For example, a nucleotide substitution that changes one codon from AAA to AAG in a gene's protein-coding region would probably not affect the protein product, because both codons specify lysine (Section 9.4). In other cases, a neutral mutation can change the amino acid sequence but not the function of a protein product.

The accumulation of neutral mutations in the DNA of a lineage can be likened to the predictable ticks of a **molecular clock**. Turn the hands of such a clock back, so the ticks wind back through the past, and the last tick will be the time when the lineage embarked on its own unique evolutionary road. To calibrate the molecular clock, the number of differences between genomes can be correlated with the timing of morphological changes observed in the fossil record.

Mutations alter the DNA of a lineage independently of all other lineages. The more recently two lineages diverged, the less time there has been for unique neutral mutations to accumulate in the DNA of each one. That is why the genomes of closely related species tend to be more similar than those of distantly related ones—a general rule that can be used to estimate relative times of divergence.

DNA and Protein Sequence Comparisons

Similarities in the amino acid sequence of a shared protein, or the nucleotide sequence of a shared gene, can be evidence of an evolutionary relationship. Biochemical comparisons like these are often used together with morphological comparisons in phylogenetic analyses, in order to provide data for hypotheses about shared ancestry.

In general, species that are more closely related tend to have more similar proteins. Two species with very few similar proteins probably have not shared an ancestor for a long time—long enough for many mutations to have accumulated in the DNA of their separate lineages. Evolutionary biologists can compare a protein's sequence among several species, and use the number of amino acid differences as one measure of relative relatedness (**FIGURE 18.6**).

The amino acids that differ are also clues. For example, a leucine to isoleucine change may not affect the function of a protein very much, because both amino acids are nonpolar, and both are about the same size. Such changes are called conservative amino acid substitutions. By contrast, the substitution of a lysine (which is basic) for an aspartic acid (which is acidic) may dramatically change the character of a protein. Such nonconservative substitutions—as well as deletions and insertions—often affect phenotype. Most mutations that affect phenotype are selected against, but occasionally one proves adaptive. Thus, the longer it has been since two lineages diverged, the more nonconservative amino acid substitutions we are likely to see when comparing their proteins.

Among species that diverged relatively recently, many proteins have identical amino acid sequences. Nucleotide sequence differences may be instructive in such cases. Even if the amino acid sequence of a protein is identical among species, the nucleotide sequence of the gene that encodes the protein may differ because of redundancies in the genetic code.

The DNA from nuclei, mitochondria, or chloroplasts can be used in nucleotide comparisons. Mitochondrial DNA accumulates mutations faster than nuclear DNA,

```
honeycreepers (10) . . . CRDVQFGWLIRNLHANGASFFFICIYLHIGRGIYYGSYLNK--ETWNIGVILLLTLMATAFVGYVLPWGQMSFWG . . .
      song sparrow . . . CRDVQFGWLIRNLHANGASFFFICIYLHIGRGIYYGSYLNK--ETWNVGIILLLALMATAFVGYVLPWGQMSFWG . . .
 Gough Island finch . . . CRDVQFGWLIRNIHANGASFFFICIYLHIGRGLYYGSYLYK--ETWNVGVILLLTLMATAFVGYVLPWGQMSFWG . . .
        deer mouse . . . CRDVNYGWLIRYMHANGASMFFICLFLHVGRGMYYGSYTFT--ETWNIGIVLLFAVMATAFMGYVLPWGQMSFWG . . .
 Asiatic black bear . . . CRDVHYGWIIRYMHANGASMFFICLFMHVGRGLYYGSYLLS--ETWNIGIILLFTVMATAFMGYVLPWGQMSFWG . . .
     bogue (a fish) . . . CRDVNYGWLIRNLHANGASFFFICIYLHIGRGLYYGSYLYK--ETWNIGVVLLLLVMGTAFVGYVLPWGQMSFWG . . .
             human . . . TRDVNYGWIIRYLHANGASMFFICLFLHIGRGLYYGSFLYS--ETWNIGIILLLATMATAFMGYVLPWGQMSFWG . . .
 thale cress (a plant) . . . MRDVEGGWLLRYMHANGASMFLIVVYLHIFRGLYHASYSSPREFVWCLGVVIFLLMIVTAFIGYVLPWGQMSFWG . . .
      baboon louse . . . ETDVMNGWMVRSIHANGASWFFIMLYSHIFRGLWVSSFTQP--LVWLSGVIILFLSMATAFLGYVLPWGQMSFWG . . .
      baker's yeast . . . MRDVHNGYILRYLHANGASFFFMVMFMHMAKGLYYGSYRSPRVTLWNVGVIIFTLTIATAFLGYCCVYGQMSHWG . . .
```

FIGURE 18.6 Example of a protein comparison. Here, part of the amino acid sequence of mitochondrial cytochrome *b* from 20 species is aligned. This protein is a crucial component of mitochondrial electron transfer chains. The honeycreeper sequence is identical in ten species of honeycreeper; amino acids that differ in the other species are shown in red. Dashes are gaps in the alignment.

FIGURE IT OUT In this comparison, which species is the most closely related to honeycreepers?   Answer: The song sparrow

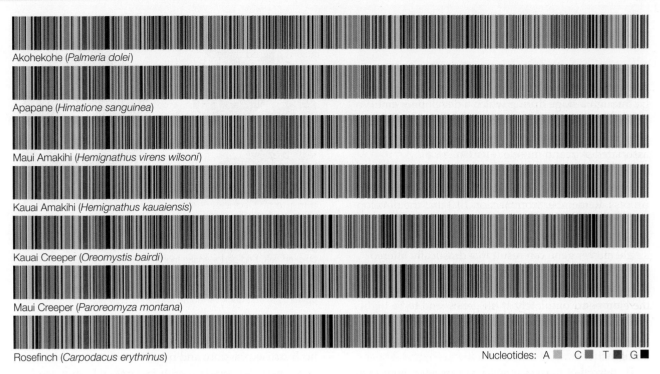

Akohekohe (*Palmeria dolei*)

Apapane (*Himatione sanguinea*)

Maui Amakihi (*Hemignathus virens wilsoni*)

Kauai Amakihi (*Hemignathus kauaiensis*)

Kauai Creeper (*Oreomystis bairdi*)

Maui Creeper (*Paroreomyza montana*)

Rosefinch (*Carpodacus erythrinus*)

Nucleotides: A ▪ C ▪ T ▪ G ▪

FIGURE 18.7 DNA barcoding. This example allows a quick visual comparison of 591 nucleotides of a mitochondrial gene called *ND2* (NADH dehydrogenase subunit 2) from 7 Hawaiian honeycreeper species and a rosefinch. The Akohekohe and Apapane are sister species, as are the Amakihis, and the Creepers.

so it can even be used to compare different individuals of the same sexually reproducing animal species. In most animals, mitochondria are inherited intact from a single parent (usually the mother). They also contain their own DNA, and they reproduce by dividing. Thus, in most cases, differences in mitochondrial DNA sequences between maternally related individuals are due to mutations, not genetic recombination events.

As you will see in the next section, some essential genes are highly conserved, which means their DNA sequences have changed very little or not at all over evolutionary time. Other genes are not conserved, and these underly differences that define species. DNA sequence differences are the basis of a method called DNA barcoding, which is used to identify an individual as belonging to a particular species. A standard region of the individual's DNA is amplified using PCR (Section 15.3) and then sequenced. The sequence is illustrated as a barcode in which each base is represented by a different color (**FIGURE 18.7**). The technique is used when other methods of identification are difficult or impossible. For example, researchers can use DNA barcoding of feces to identify components of an endangered animal's diet.

Getting useful information from comparing DNA requires a lot more data than comparing proteins. This is because coincidental homologies are statistically more likely to occur with DNA comparisons—there are only four nucleotides in DNA versus twenty amino acids in proteins. However, DNA sequencing has become so fast that there is a lot of data available to compare. Genomics studies with such data have shown us (for example) that about 88 percent of the mouse genome sequence is identical with the human genome, as is 73 percent of the zebrafish genome, 47 percent of the fruit fly genome, and 25 percent of the rice genome.

> **TAKE-HOME MESSAGE 18.4**
>
> **Why do DNA or protein similarities reflect evolutionary history?**
>
> ✔ Mutations change the nucleotide sequence of a lineage's DNA over time.
>
> ✔ Lineages that diverged long ago generally have more differences between their DNA (and their proteins) than do lineages that diverged more recently.

molecular clock Technique that uses molecular change to estimate how long ago two lineages diverged.

✔ Similar patterns of embryonic development are an outcome of highly conserved master genes.

In general, the more closely related animals are, the more similar is their development. For example, all vertebrates go through a stage during which a developing embryo has four limb buds, a tail, and a series of somites—divisions of the body that give rise to the backbone and associated skin and muscle (**FIGURE 18.8**).

Animals have similar patterns of embryonic development because the very same master genes direct the process. Remember from Section 10.3 that the development of an embryo into a body is orchestrated by layer after layer of master gene expression. The failure of any single master gene can result in a drastically altered body plan, typically with devastating consequences. Because a mutation in a master gene typically unravels development completely, these genes tend to be highly conserved. Even among lineages that diverged a very long time ago, such genes often retain similar sequences and functions.

If conserved master genes direct development in all vertebrate lineages, how do the adult forms end up so different? Part of the answer is that there are differences in the onset, rate, or completion of early steps in development. These differences are brought about by variations in master gene expression patterns. Consider homeotic genes called *Hox*. Like other homeotic genes, *Hox* gene expression helps sculpt details of the body's form during embryonic development. Vertebrate animals have multiple sets of the same ten *Hox* genes that occur in insects and other arthropods. You have already read about one of these genes, *antennapedia*, which determines the identity of the thorax (the body part with legs) in fruit flies. One vertebrate version of *antennapedia* is called *Hoxc6*, and it determines the identity of the back (as opposed to the neck or tail). Expression of the *Hoxc6* gene causes ribs to develop on a vertebra.

FIGURE 18.9 How differences in body form can arise from differences in master gene expression. Expression of the *Hoxc6* gene is indicated by purple stain in two vertebrate embryos, chicken (left) and garter snake (right). Expression of this gene causes a vertebra to develop ribs. Chickens have 7 vertebrae in their back and 14 to 17 vertebrae in their neck; snakes have upwards of 450 back vertebrae and essentially no neck.

Vertebrae of the neck and tail normally develop with no *Hoxc6* expression, and no ribs (**FIGURE 18.9**). *Hox* genes also regulate limb formation. Body appendages as diverse as crab legs, beetle legs, sea star arms, butterfly wings, fish fins, and mouse feet start out as clusters of cells that bud from the surface of the embryo. The buds form wherever a homeotic gene called *Dlx* is expressed. *Dlx* encodes a transcription factor that signals clusters of embryonic cells to "stick out from the body" and give rise to an appendage. *Hox* genes suppress *Dlx* expression in all parts of an embryo that will not have appendages.

TAKE-HOME MESSAGE 18.5
Why do animals develop in similar ways?

✔ Similarities in patterns of development are the result of master genes that have been conserved over evolutionary time.

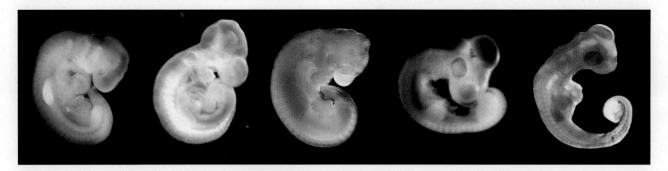

FIGURE 18.8 Visual comparison of vertebrate embryos. All vertebrates go through an embryonic stage in which they have four limb buds, a tail, and divisions called somites along their back. From left to right: human, mouse, bat, chicken, alligator.

CREDITS: (8) from left, © Lennart Nilsson/Bonnierforlagen AB; Courtesy of Anna Bigas, IDIBELL-Institut de Recerca Oncologica, Spain; From *Embryonic staging system for the short-tailed fruit bat, Carollia perspicillata, a model organism for the mammalian order Chiroptera, based upon timed pregnancies in captive-bred animals.* C.J. Cretekos et al., Developmental Dynamics Volume 233, Issue 3, July 2005, Pages: 721–738. Reprinted with permission of Wiley-Liss, Inc. a subsidiary of John Wiley & Sons, Inc.; Courtesy of Prof. Dr. G. Elisabeth Pollerberg, Institut für Zoologie, Universität Heidelberg, Germany; USGS; (9) Courtesy of Ann C. Burke, Wesleyan University.

✔ We use phylogeny to understand how to preserve species that exist today.

Studies of phylogeny reveal how species relate to one another and to species that are now extinct. In doing so, they inform our understanding of how shared ancestry interconnects all species—including our own.

Conservation Biology

The story of the Hawaiian honeycreepers is a dramatic illustration of how evolution works. It also shows how finding ancestral connections can help species that are still living. As more and more honeycreeper species become extinct, the group's reservoir of genetic diversity dwindles. The lowered diversity means the group as a whole is less resilient to change, and more likely to suffer catastrophic losses.

Deciphering their phylogeny can tell us which honeycreeper species are most different from the others—and those are the ones most valuable in terms of preserving the group's genetic diversity. Such research allows us to concentrate our resources and conservation efforts on those species whose extinction would mean a greater loss to biodiversity. For example, we now know the Poouli (pictured in FIGURE 18.1C) to be the most distant relative in the Hawaiian honeycreeper family. Unfortunately, the knowledge came too late; the Poouli is probably extinct. Its extinction means the loss of a large part of evolutionary history of the group: One of the longest branches of the honeycreeper family tree is gone forever.

Cladistics analyses are also used to correlate past evolutionary divergences with behavior and dispersal patterns of existing populations. Such studies are useful in conservation efforts. For example, a decline in antelope populations in African savannas is at least partly due to competition with domestic cattle. A cladistic analysis of mitochondrial DNA sequences suggested that current populations of blue wildebeest (FIGURE 18.10) are genetically less similar than they should be, based on other antelope groups of similar age. Combined with behavioral and geographic data, the analysis helped conservation biologists realize that a patchy distribution of preferred food plants is preventing gene flow among blue wildebeest populations. The absence of gene flow can lead to a catastrophic loss of genetic diversity in populations under pressure. Restoring appropriate grasses in intervening, unoccu-

FIGURE 18.10 A blue wildebeest in Africa. Conservation biologists discovered that a patchy availability of preferred food was hampering gene flow among wildebeest populations. The biologists recommended restoring grasses in some areas that had been cleared, to re-establish gene flow among isolated wildebeests.

pied areas of savanna would allow isolated wildebeest populations to reconnect.

Medical Research

Researchers often study the evolution of viruses and other infectious agents by grouping them into clades based on biochemical characters. Even though viruses are not alive, they can mutate every time they infect a host, so their genetic material changes over time. Consider the H5N1 strain of influenza (flu) virus, which infects birds and other animals. H5N1 has a very high mortality rate in humans, but human-to-human transmission has been rare to date. However, the virus replicates in pigs without causing symptoms. Pigs transmit the virus to other pigs—and apparently to humans too. A phylogenetic analysis of H5N1 isolated from pigs showed that the virus "jumped" from birds to pigs at least three times since 2005, and that one of the isolates had acquired the potential to be transmitted among humans. An increased understanding of how this virus adapts to new hosts is helping researchers design more effective vaccines for it.

> **TAKE-HOME MESSAGE 18.6**
> **How can we use what we learn about evolutionary history?**
>
> ✔ Phylogeny research is yielding an ever more specific and accurate picture of how all life is related by shared ancestry.
>
> ✔ Among other applications, phylogeny research can help us to prioritize efforts to preserve endangered species, and to understand the spread of infectious diseases.

CREDITS: (in text) John Steiner/Smithsonian Institution; (10) Alan Lucas/Shutterstock.

CHAPTER 18 **301**
ORGANIZING INFORMATION ABOUT SPECIES

Bye Bye Birdie (revisited)

In 2004, researchers captured one of the three remaining Pooulis, with the intent of starting a captive breeding program before the species became extinct. They were unable to capture a female to mate with this male before he died in captivity a month later. Cells from this last bird were frozen, and may be used in the future for cloning. However, with no parents left to demonstrate the species' natural behavior to chicks, cloned birds would probably never be able to establish themselves as a natural population.

summary

Section 18.1 The Hawaiian honeycreepers are highly specialized. Their specialization makes them particularly vulnerable to extinction as a result of habitat loss and exotic species introductions that followed human colonization of the Hawaiian Islands.

Section 18.2 Evolutionary biologists reconstruct evolutionary history (**phylogeny**) by comparing physical, behavioral, and biochemical traits, or **characters**, among species. A **clade** is a **monophyletic group** that consists of an ancestor in which one or more **derived traits** evolved, together with all of its descendants.

Making hypotheses about the evolutionary history of a group of clades is called **cladistics. Evolutionary tree** diagrams are based on the premise that all organisms are connected by shared ancestry. In a **cladogram**, each line represents a lineage. A lineage branches into two **sister groups** at a node, which represents a shared ancestor.

Section 18.3 Comparative morphology is one way to study evolutionary connections among lineages. **Homologous structures** are similar body parts that became modified differently in different lineages (a pattern called **morphological divergence**). Such parts are evidence of a common ancestor. **Analogous structures** are body parts that look alike in different lineages but did not evolve in a common ancestor. Rather, they evolved separately after the lineages diverged, a pattern called **morphological convergence**.

Section 18.4 We can discover and clarify evolutionary relationships through comparisons of DNA and protein sequences. In general, these sequences are more similar among lineages that diverged more recently. Neutral mutations accumulate in DNA at a predictable rate; like the ticks of a **molecular clock**, they help researchers estimate how long ago lineages diverged.

Section 18.5 Master genes that affect development tend to be highly conserved, so similarities in patterns of embryonic development reflect shared ancestry that can be evolutionarily ancient. Mutations that alter the timing of master gene expression can alter the rate or onset of development, and thus result in different adult body forms.

Section 18.6 Reconstructing phylogeny is part of our efforts to preserve endangered species. Phylogeny is also used for studying the spread of viruses and other agents of infectious diseases.

self-quiz

1. In cladistics, the only taxon that is always correct as a clade is the _____ .
   - a. genus
   - b. family
   - c. species
   - d. kingdom

2. In evolutionary trees, each node represents a(n) _____ .
   - a. single lineage
   - b. extinction
   - c. point of divergence
   - d. adaptive radiation

3. A clade is _____ .
   - a. defined by a derived trait
   - b. a monophyletic group
   - c. a hypothesis
   - d. all of the above

4. Cladistics _____ .
   - a. may involve parsimony analysis
   - b. is based on derived traits
   - c. both of the above are correct

5. In cladograms, sister groups are _____ .
   - a. inbred
   - b. the same age
   - c. represented by nodes
   - d. in the same family

6. Through _____ , a body part of an ancestor is modified differently in different lines of descent.
   - a. homologous evolution
   - b. morphological convergence
   - c. adaptive divergence
   - d. morphological divergence

7. Homologous structures among major groups of organisms may differ in _____ .
   - a. size
   - b. shape
   - c. function
   - d. all of the above

8. Neutral mutations are those that do not affect _____ .
   - a. amino acid sequence
   - b. nucleotide sequence
   - c. the chances of survival
   - d. all of the above

9. Mitochondrial DNA sequences are often used in cladistic comparisons of _____ .
   - a. different species
   - b. individuals of the same species
   - c. different taxa

Hawaiian Honeycreeper Phylogeny The Poouli (*Melamprosops phaeosoma*) was discovered in 1973 by a group of students from the University of Hawaii. Its membership in the Hawaiian honeycreeper clade was (until recently) controversial, mainly because its appearance and behavior are so different from other living honeycreepers. It particularly lacked the "old tent" odor characteristic of other honeycreepers.

In 2011, Heather Lerner and her colleagues deciphered phylogeny of the 19 Hawaiian honeycreepers that were not yet officially declared to be extinct at the time, including the Poouli. The researchers sequenced mitochondrial and nuclear DNA samples taken from the honeycreepers, and also from 28 other birds (outgroups). Phylogenetic analysis of these data firmly establishes the Poouli as a member of the clade, and also reveals the Eurasian rosefinch as the clade's closest relative (**FIGURE 18.11**).

1. Which species on the cladogram represents an outgroup?

2. Which species is most closely related to the Apapane (*Himatione sanguinea*)?

3. What is the sister group of the Akikiki (*Oreomystis bairdi*)?

4. Which species is more closely related to the Palila (*Loxioides bailleui*): the Iiwi (*Vestiaria coccinea*) or the Maui Alauahio (*Paroreomyza montana*)?

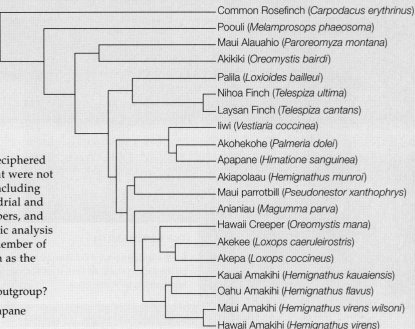

- Common Rosefinch (*Carpodacus erythrinus*)
- Poouli (*Melamprosops phaeosoma*)
- Maui Alauahio (*Paroreomyza montana*)
- Akikiki (*Oreomystis bairdi*)
- Palila (*Loxioides bailleui*)
- Nihoa Finch (*Telespiza ultima*)
- Laysan Finch (*Telespiza cantans*)
- Iiwi (*Vestiaria coccinea*)
- Akohekohe (*Palmeria dolei*)
- Apapane (*Himatione sanguinea*)
- Akiapolaau (*Hemignathus munroi*)
- Maui parrotbill (*Pseudonestor xanthophrys*)
- Anianiau (*Magumma parva*)
- Hawaii Creeper (*Oreomystis mana*)
- Akekee (*Loxops caeruleirostris*)
- Akepa (*Loxops coccineus*)
- Kauai Amakihi (*Hemignathus kauaiensis*)
- Oahu Amakihi (*Hemignathus flavus*)
- Maui Amakihi (*Hemignathus virens wilsoni*)
- Hawaii Amakihi (*Hemignathus virens*)

FIGURE 18.11 Hawaiian honeycreeper phylogeny. This cladogram was constructed using sequence comparisons of mitochondrial DNA (whole genome), and 13 nuclear DNA loci of 19 Hawaiian honeycreepers and 28 other finch species.

10. Molecular clocks are based on comparisons of the number of _____ mutations between species.
    a. lethal
    b. neutral
    c. conservative
    d. nonconservative

11. True or false? DNA barcoding can identify an individual as belonging to a particular species.

12. A mutation that alters the embryonic expression pattern of a _____ may lead to major differences in the adult form.
    a. derived trait
    b. master gene
    c. homologous structure
    d. all of the above

13. All of the following data types can be used as evidence of shared ancestry except similarities in _____ .
    a. amino acid sequences
    b. DNA sequences
    c. fossil morphologies
    d. embryonic development
    e. form due to convergence
    f. all are appropriate

14. True or false? Phylogeny helps us study the spread of viruses through human populations.

15. Match the terms with the most suitable description.
    ___ phylogeny
    ___ cladogram
    ___ homeotic genes
    ___ homologous structures
    ___ molecular clock
    ___ analogous structures
    ___ derived trait

    a. novel character
    b. evolutionary history
    c. human arm and bird wing
    d. similar across diverse taxa
    e. measures neutral mutations
    f. insect wing and bird wing
    g. evolutionary tree

critical thinking

1. In the late 1800s, a biologist studying animal embryos coined the phrase "ontogeny recapitulates phylogeny," meaning that the physical development of an animal embryo (ontogeny) seemed to retrace the changing form of the species during its evolutionary history (phylogeny). Why would embryonic development retrace evolutionary steps?

2. The photos shown below illustrate a case of synpolydactyly. This genetic abnormality is characterized by webbing between partially or completely duplicated fingers or toes. The same mutations that give rise to the human phenotype also give rise to a similar phenotype in mice. In which family of genes do you think these mutations occur?

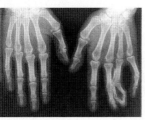

CENGAGE brain.com To access course materials, please visit www.cengagebrain.com.

CREDITS: (11) © Cengage Learning; (in text CT #3) Courtesy of Dr. Sajid Malik, from www.biomedcentral.com/1471-2350/8/78.

19 Life's Origin and Early Evolution

LEARNING ROADMAP

This chapter explains how the molecular subunits of life introduced in Section 3.3 could have originated and assembled to form the first cells. It considers the evolution of eukaryotic traits (Sections 4.5–4.12) and draws on your understanding of fossils and the geologic time scale (16.5 and 16.8), aerobic respiration (7.2), and photosynthesis (6.4).

THE EARLY EARTH

Earth formed about 4.6 billion years ago from material released by exploding stars. At first, it lacked oxygen and was constantly bombarded by meteorites.

BUILDING BLOCKS OF LIFE

Simulations show how organic monomers could have formed on the early Earth. Such compounds also form in space and could have been delivered by meteorites.

ORIGIN OF CELLS

Experiments provide insight into how cells could have arisen from nonliving material through known physical and chemical processes.

EVOLUTION OF EARLY LIFE

The first cells were probably anaerobic and prokaryotic. Evolution of oxygen-producing photosynthesis altered Earth's atmosphere, creating selection pressure that favored aerobic organisms.

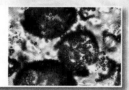

ORIGINS OF EUKARYOTIC ORGANELLES

The nucleus and endomembrane system are evolutionarily derived from infoldings of the plasma membrane. Mitochondria and chloroplasts are thought to be descendants of bacteria.

We continue our discussion of the features of bacterial and archaeal cells in Sections 20.5–20.8. You will learn much more about the features of simple eukaryotes when we discuss the protists in Chapter 21. We return to the topic of the ozone layer in Section 48.6 as we discuss the damage done by air pollutants.

We live in a vast universe that we have only begun to explore. So far, we know of only one planet—Earth—that has life. In addition, biochemical, genetic, and metabolic similarities among Earth's species imply that all evolved from a common ancestor that lived billions of years ago. What properties of the ancient Earth allowed life to arise, survive, and diversify? Could similar processes occur on other planets? These are some of the questions posed by **astrobiology**, the study of life's origins and distribution in the universe.

Astrobiologists study Earth's extreme habitats to determine the range of conditions that can support life. They have found species that withstand extraordinary levels of temperature, pH, salinity, and pressure. For example, some bacteria live deep in the soil of Chile's Atacama Desert, the driest place on Earth (**FIGURE 19.1**). Other bacteria thrive at high pressure and temperature 3 kilometers (almost 2 miles) beneath the soil surface in Virginia. There is even a biofilm (Section 4.4) under Antarctica's western ice sheet.

Understanding how these species withstand such extreme conditions on Earth informs the search for life elsewhere. For example, the details of metabolism vary, but in all known species—even the extraordinary ones—metabolic reactions occur in or at the boundary of water-based solutions. Thus, liquid water is considered an essential requirement for life as we know it. Excitement over the discovery of water on Mars, our closest planetary neighbor, stems from this assumption. A robotic lander found water frozen in the soil of Mars, and geological formations suggest water flowed across the planet's surface and pooled in giant lakes billions of years ago. Other features suggest a small amount of water may still flow seasonally in some areas.

If there is any life on Mars today, it is likely to be underground. Unlike Earth, Mars does not have an **ozone layer**: an atmospheric layer with a high concentration of ozone (O_3). Earth's ozone layer serves as a natural sunscreen, preventing most ultraviolet (UV) radiation emitted by the sun from reaching the planet's surface. Because Mars has no ozone layer, its surface receives intense UV radiation. As explained in Section 8.6, UV radiation damages DNA. Thus, the UV radiation that reaches Mars probably sterilizes the upper layer of the Martian soil.

Suppose scientists do find evidence that microbial life exists or existed on Mars, or on some other planet. Why would it matter? The discovery of extraterrestrial microbes would lend credence to the idea that nonhuman intelligent life exists somewhere else in the universe. The more places microbial life exists, the more likely it is that complex, intelligent life evolved on planets other than Earth.

This chapter is your introduction to a slice through time. We begin with Earth's formation and move on to life's chemical origins and the evolution of traits present in modern eukaryotes. The picture we paint here sets the stage for the next unit, which takes you along lines of descent to the present range of biodiversity.

astrobiology The scientific study of life's origin and distribution in the universe.

ozone layer High atmospheric layer rich in ozone; prevents most ultraviolet radiation in sunlight from reaching Earth's surface.

FIGURE 19.1 Looking for life in a part of Chile's Atacama Desert, where rain falls only once every few decades and the soil evolved without plant life. Scientist Jay Quade of the University of Arizona was a member of a team that found bacteria living 30 centimeters beneath the surface of this arid desert soil.

CREDITS: (opposite) Michael Aw/Lonely Planet Images/Getty Images; (1) Photo by Julio Betancourt/U.S. Geological Survey. See Drees, K.P. et al. 2006. Bacterial community structure of soils in a hyperarid region of the Atacama Desert. Applied and Environmental Microbiology 72, 7902-7908 and Quade, J et al, 2007, Soils at the hyperarid margin: the isotopic composition of soil carbonate from the Atacama Desert. Geochimica et Cosmochimica Acta 71, 3772-3795

FIGURE 19.2 Artist's depiction of our sun surrounded by a cloud of dust, debris, and gases. Earth and other planets formed from the material in this cloud.

✔ Modern chemistry and physics are the bases for scientific hypotheses about early events in Earth's history.

Origin of the Universe and Earth

Studies of the modern universe allow astronomers and physicists to propose and test ideas about its origin. According to the **big bang theory**, the universe began in a single instant, about 13 to 15 billion years ago. In that instant, all existing matter and energy suddenly appeared and exploded outward from a single point. Simple elements such as hydrogen and helium formed within minutes. Then, over millions of years, gravity drew the gases together and they condensed to form giant stars.

Explosions of these early stars scattered heavier elements from which today's galaxies formed. About 5 billion years ago, a cloud of dust and rocks (asteroids) orbited the star we now call our sun (**FIGURE 19.2**). The asteroids collided and merged into bigger asteroids. The heavier these pre-planetary objects became, the more gravitational pull they exerted, and the more material they gathered. By about 4.6 billion years ago, this gradual buildup of materials had formed Earth and the other planets of our solar system.

Conditions on the Early Earth

Planet formation did not remove all the material orbiting the sun, so the early Earth received a constant hail of meteorites and was struck by many asteroids. This extraterrestrial material, along with substances released by frequent volcanic eruptions, provided components of Earth's land, seas, and atmosphere.

The composition of Earth's early atmosphere remains a matter of debate, but geologic evidence suggests our planet started out with little or no free oxygen (O_2). Had O_2 been present early on, we would see evidence of iron oxidation (rust formation) in Earth's most ancient rocks. However, these rocks show no sign of such oxidation. The apparent lack of O_2 interests scientists because it would have facilitated some proposed steps on the path to life. Had O_2 been present, oxidation reactions would have broken apart small organic compounds as quickly as they formed.

Liquid water is essential to life as we know it because molecules that carry out metabolic reactions have to be dissolved in water. At first, Earth's surface was molten rock, so all water was in the form of vapor. However, evidence from ancient rocks indicates that by 4.3 billion years ago, Earth had cooled enough for water to pool on its surface.

big bang theory Well-supported hypothesis that the universe originated by a nearly instant distribution of matter through space.

> **TAKE-HOME MESSAGE 19.2**
> **What were conditions like on the early Earth?**
>
> ✔ During Earth's early years, meteorites pummeled the planet's surface and volcanic eruptions were common.
>
> ✔ The early atmosphere probably had no oxygen gas.
>
> ✔ As Earth cooled, liquid water began to pool on its surface.

CREDIT: (2) Painting by William K. Hartmann.

19.3 Organic Monomers Form

✔ All living things are made from the same organic subunits: amino acids, fatty acids, nucleotides, and simple sugars.

Organic Molecules From Inorganic Precursors

Until the early 1900s, chemists thought that organic molecules possessed a special "vital force" and that only living organisms could make them. Then, in 1925, a chemist synthesized urea, an organic molecule abundant in urine. Later, another chemist synthesized alanine, an amino acid. These synthesis reactions showed organic molecules could be formed by nonliving processes.

Sources of Life's First Building Blocks

Today, there are three main hypotheses regarding the mechanism by which organic monomers appeared on the early Earth. Keep in mind that these mechanisms are not mutually exclusive. All three may have operated simultaneously and contributed to an accumulation of simple organic compounds in Earth's early seas.

Lightning-Fueled Atmospheric Reactions

In 1953, Stanley Miller and Harold Urey tested the hypothesis that lightning could have powered synthesis reactions in Earth's early atmosphere. To simulate this process, they filled a reaction chamber with methane, ammonia, and hydrogen gas, and zapped it with sparks from electrodes (**FIGURE 19.3**). Within a week, a variety of organic molecules formed, including amino acids that are common in living things. Our understanding of the

FIGURE 19.4 A hydrothermal vent on the seafloor. Mineral-rich water heated by geothermal energy streams out, into cold ocean water. As the water cools, dissolved minerals come out of solution and form a chimney-like structure around the vent.

composition of Earth's early atmosphere has changed since that time, but more recent experimental simulations using gas mixtures that more accurately represent the early atmosphere also produced amino acids.

Reactions at Hydrothermal Vents By another hypothesis, synthesis of life's building blocks occurred at deep-sea hydrothermal vents. A **hydrothermal vent** is like an underwater geyser, a place where mineral-rich water heated by geothermal energy streams out through a rocky opening in the seafloor (**FIGURE 19.4**).

Delivery From Space Modern-day meteorites that fall to Earth sometimes contain amino acids, sugars, and nucleotide bases and these compounds (or precursors of them) have been discovered in gas clouds surrounding nearby stars. Thus it is possible that some of the many meteorites that fell on the early Earth carried organic monomers that had formed in outer space.

---

hydrothermal vent Underwater opening from which mineral-rich water heated by geothermal energy streams out.

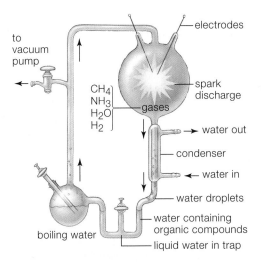

FIGURE 19.3 ▶**Animated** Diagram of an apparatus used to test whether organic compounds could have formed spontaneously on the early Earth. Water vapor, hydrogen gas (H_2), methane (CH_4), and ammonia (NH_3) simulated Earth's early atmosphere. Sparks from an electrode simulated lightning.

FIGURE IT OUT What was the source of the nitrogen for the amino acids formed in this apparatus?

Answer: Ammonia

CREDITS: (3) © Cengage Learning; (4) Courtesy of the University of Washington.

TAKE-HOME MESSAGE 19.3

What was the source of the small organic molecules required to build the first life?

✔ Small organic molecules that serve as the building blocks for living things can be formed by nonliving mechanisms. For example, amino acids form in reaction chambers that simulate conditions on the early Earth. They are also present in some meteorites.

19.4 From Polymers to Protocells

✔ We will never know for sure how the first cells came to be, but we can investigate the possible steps on the road to life.

Properties of Cells

In addition to sharing the same molecular components, all cells have a plasma membrane with a lipid bilayer. They have a genome of DNA that enzymes transcribe into RNA, and ribosomes that translate RNA into proteins. All cells replicate, and they pass on copies of their genetic material to their descendants. The many similarities in structure, metabolism, and replication processes among all life provide evidence of descent from a common cellular ancestor.

No one was around to witness the origin of life on Earth, and time has erased all traces of the earliest cells. However, scientists can still investigate this first chapter in life's history by making hypotheses about how life began, and testing those hypotheses experimentally. Their work has shown that cells could have arisen in a stepwise process beginning with inorganic materials (**FIGURE 19.5**). Each step on this hypothetical road to life can be explained by familiar chemical and physical mechanisms that still occur today.

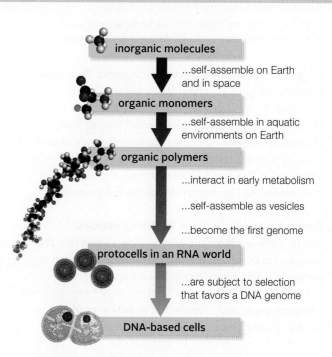

FIGURE 19.5 ▶Animated Proposed sequence for the evolution of cells. Scientists carry out experiments and simulations that test the feasibility of individual steps.

Origin of Metabolism

Modern cells take up organic monomers, concentrate them, and assemble them into organic polymers. Before there were cells, a nonbiological process that concentrated organic subunits would have increased the chance of polymer formation. By one hypothesis, this process occurred on clay-rich tidal flats. Clay particles have a slight negative charge, so positively charged molecules (such as organic subunits dissolved in seawater) stick to them. Such binding would have concentrated the subunits. Evaporation at low tide would have continued this process, and energy from the sun might have triggered polymerization. Amino acids do form short chains under simulated tidal flat conditions.

The **iron–sulfur world hypothesis**, proposed by Günter Wächtershäuser, holds that life originated at deep-sea hydrothermal vents. The porous rocks that form around these vents are rich in iron sulfides, compounds that easily donate electrons to inorganic gases dissolved in the hot seawater spewing from the vents. Accepting electrons causes the gases (such as carbon dioxide and hydrogen cyanide) to react and form organic molecules (such as pyruvate). The organic molecules stick to the iron sulfur compounds and accumulate inside tiny, cell-sized chambers of the rocks, where they can undergo further reactions as they become concentrated. Long ago, such molecules became catalytic: early versions

of enzymes that carry out modern metabolic reactions. The iron sulfur cofactors required by all modern organisms may be a legacy of life's rocky beginnings at hydrothermal vents.

Origin of the Genome

All modern cells have a genome of DNA. They pass copies of their DNA to descendant cells, which use instructions encoded in the DNA to build proteins. Some of these proteins are enzymes that synthesize new DNA, which is passed along to descendant cells, and so on. Thus, protein synthesis depends on DNA, which is built by proteins. How did this cycle begin?

In the 1960s, Francis Crick and Leslie Orgel addressed this dilemma by suggesting that RNA may have been the first molecule to encode genetic information, a concept known as the **RNA world hypothesis**. Evidence that RNA can both store genetic information and function like an enzyme in protein synthesis, supports this hypothesis. **Ribozymes**, or RNAs that function as enzymes, are common in living cells. For example, the rRNA in ribosomes speeds formation of peptide bonds during protein synthesis. Other ribozymes cut noncoding bits (introns) out of newly formed mRNAs. Researchers have also produced self-replicating ribozymes that assemble free nucleotides.

CREDIT: (5) © Cengage Learning.

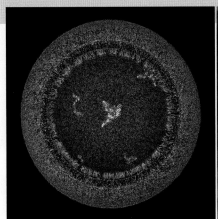

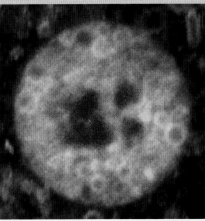

A Illustration of a laboratory-produced protocell with a bilayer membrane of fatty acids and strands of RNA inside.

B Laboratory-formed protocell consisting of RNA-coated clay (red) surrounded by fatty acids and alcohols.

C Field-testing a hypothesis about protocell formation. David Deamer pours a mix of small organic molecules and phosphates into a hot acidic pool in Russia.

FIGURE 19.6 ▶**Animated** Protocells. Scientists test hypotheses about protocell formation through laboratory simulations and field experiments.

If the earliest self-replicating genetic systems were RNA-based, then why do all organisms now have a genome of DNA? The structure of DNA may hold the answer. Compared to a double-stranded DNA molecule, single-stranded RNA breaks more easily. Thus, a switch from RNA to DNA would make larger, more stable genomes possible.

Origin of the Plasma Membrane

Self-replicating molecules and products of other early synthetic reactions would have floated away from one another unless something enclosed them. In modern cells, a plasma membrane serves this function. If the first reactions took place in tiny rock chambers, the rock would have acted as a boundary. Over time, lipids produced by reactions inside a chamber could have accumulated and lined the chamber wall, forming a protocell. A **protocell** is a membrane-enclosed collection of interacting molecules that can take up material and replicate. Protocells are thought to be the ancestors of cellular life.

Experiments by Jack Szostak and others have shown that protocells can form even without rock chambers. **FIGURE 19.6A** illustrates one type of protocell that Szostak investigates. **FIGURE 19.6B** is a photo of a protocell that formed in his laboratory. This synthetic protocell consists of a membrane of lipid bilayer enclosing strands of RNA. The protocell can "grow" by adding fatty acids to its membrane and nucleotides to its RNA; mechanical force can make it "divide."

David Deamer studies protocell formation both in the laboratory and in the field. In the lab, he has shown that the small organic molecules carried to Earth on meteorites can react with minerals and seawater to form vesicles with a bilayer membrane. However, Deamer has yet to locate a natural environment that facilitates the same process. In one experiment, he added a mix of organic subunits to the acidic waters of a clay-rich volcanic pool in Russia (**FIGURE 19.6C**). The organic subunits bound tightly to the clay, but no vesicle-like structures formed. Deamer concluded that hot acidic waters of volcanic springs do not provide the right conditions for protocell formation. He continues to carry out experiments to determine what naturally occurring conditions would favor this process.

iron–sulfur world hypothesis Hypothesis that the metabolic reactions that led to the first cells took place on the porous surface of iron–sulfide-rich rocks at hydrothermal vents.
protocell Membranous sac that contains interacting organic molecules; hypothesized to have formed prior to the earliest life-forms.
ribozyme RNA that functions as an enzyme.
RNA world hypothesis Hypothesis that RNA served as the genetic information of early life.

> ## TAKE-HOME MESSAGE 19.4
> ### What have existing life and simulations revealed about the steps that led to the first cells?
>
> ✔ All living cells carry out metabolic reactions, are enclosed within a plasma membrane, and can replicate themselves.
>
> ✔ Metabolic reactions may have begun when molecules became concentrated on clay particles or inside tiny rock chambers near hydrothermal vents.
>
> ✔ RNA can serve as an enzyme, as well as a genome. An RNA world may have preceded evolution of DNA-based genomes.
>
> ✔ Vesicle-like structures with outer membranes form spontaneously when certain organic molecules are mixed with water.

CREDITS: (6A) © Janet Iwasa; (6B) From Hanczyc, Fujikawa, and Szostak, Experimental Models of Primitive Cellular Compartments: Encapsulation, Growth, and Division; www.sciencemag.org, Science 24 October 2003; 302;529, Figure 2, Page 619. Reprinted with permission of the authors and AAAS; (6C) Photo by Tony Hoffman, courtesy of David Deamer.

✔ Fossils and molecular comparisons among modern organisms inform us about the early history of life.

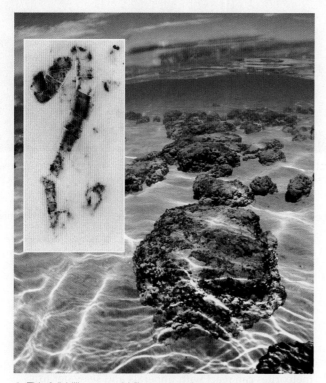

A This 3.5-billion-year-old filament may be a chain of fossil bacteria from an ancient stromatolite. The underlying photo shows modern stromatolites in Australia's Shark Bay. Each stromatolite consists of living photosynthetic bacteria atop the remains of countless earlier generations of cells along with the sediment that they trapped.

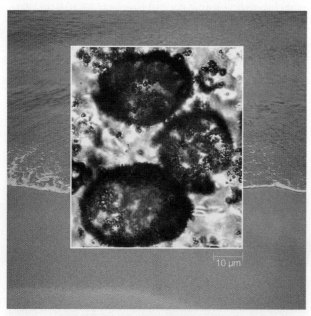

B These 3.4-billion-year-old spheres may be fossil bacteria that lived among sand grains of an ancient shore.

FIGURE 19.7 The oldest proposed bacterial microfossils.

The Common Ancestor of All Life

The processes described in Section 19.4 may have produced cellular life more than once. If so, all but one of those early cell lineages became extinct. Analysis of modern genomes tells us that all living species are descended from the same cell, and that this cell may have lived as early as 4 billion years ago.

Given what scientists know about relationships among modern species, most assume that this common ancestor was prokaryotic, meaning it did not have a nucleus (Section 1.4). Oxygen was scarce on the early Earth, so the ancestral cell must also have been anaerobic (capable of living without oxygen).

Looking for Evidence of Early Life

Finding and identifying signs of early cells poses a challenge. Cells are microscopic and most have no hard parts to fossilize. In addition, few ancient rocks that could hold early fossils still exist. Tectonic plate movements have destroyed nearly all rocks older than about 4 billion years, and most slightly younger rocks have been heated or undergone other processes that destroy traces of biological material. To add to the difficulty, structures formed by nonbiological mechanisms sometimes resemble fossils. To avoid mistakenly accepting such material as a genuine fossil, scientists constantly reanalyze purported fossil finds and they often question one another's conclusions.

The Oldest Fossil Cells

The divergence that separated the two prokaryotic domains, Bacteria and Archaea, occurred very early in the history of life, and no fossils from before this divergence have been discovered. At present, the oldest proposed cell microfossils (microscopic fossils) are filaments from 3.5-billion-year-old rocks in Western Australia. The filaments resemble chains of modern photosynthetic bacteria, and the rocks in which they occur are thought to be remains of ancient stromatolites (**FIGURE 19.7A**).

A **stromatolite** is a mounded, layered structure that forms in shallow sunlit water when a mat of photosynthetic bacteria traps minerals and sediment. The stromatolite increases in size over time as new layers form over the old. When sediment covers the living bacteria atop the stromatolite, new bacterial cells grow over that sediment, then trap more sediment to form a new layer. Scientists can observe this process in modern stromatolites such as those in Australia's Shark Bay. Stromatolites in this bay began growing about 2,000 years ago and now stand up to 1.5 meters high.

CREDITS: (7A) background, Michael Aw/Lonely Planet Images/Getty Images; inset, Courtesy of John Fuerst, University of Queensland. Originally published in *Archives of Microbiology* vol 175, p 413–29 (Lindsay MR, Webb RI, Strous M, Jetten MS, Butler MK, Forde RJ, Fuerst JA. Cell compartmentalisation in planctomycetes: Novel types of structural organization for the bacterial cell. *Arch. Microbiol.* 2001 Jun, 175(6): 413–29); (7B) background, tratong/Shutterstock.com; inset, Courtesy of David Wacey.

Stromatolites reached their peak abundance about 1.25 billion years ago, when they were common worldwide.

Another set of proposed microfossils, also from Western Australia, dates to 3.4 billion years ago. The fossils have a spherical shape and occur in rocks composed of sand grains from an ancient beach (**FIGURE 19.7B**). The presence of pyrite (an iron–sulfide mineral) in these fossils suggests they were similar to the sulfur bacteria common in modern mud flats. Sulfur bacteria use sulfur as the final electron acceptor in their energy-producing pathway, and they produce pyrite as a by-product of this process.

Changes in the Air

Many types of bacteria carry out photosynthesis, but only one group, the cyanobacteria, do so by an oxygen-producing pathway. Evolution of oxygen-producing photosynthesis in cyanobacteria had a dramatic effect on early life. By about 2.5 billion years ago, oxygen released by these bacteria had begun to accumulate in Earth's air and seas creating a new, global selection pressure. The oxygen was toxic to many species that had evolved in its absence. Inside cells, it reacts with metal ions, forming free radicals that can damage DNA and other essential cell components (Sections 2.3 and 5.6). Species unable to detoxify the free radicals either went extinct or became restricted to the low-oxygen environments that remained. By contrast, species that happened to have metabolic machinery capable of detoxifying the free radicals thrived. Some of these survivors even began to use some of this machinery for aerobic respiration, an energy-releasing pathway in which oxygen serves as the final electron acceptor (Section 7.5).

The increase in atmospheric oxygen also led to the formation of the ozone layer, a region of the upper atmosphere that contains a high concentration of ozone gas (O_3). Before the ozone layer formed, life could exist only in water, which shielded organisms from incoming UV radiation. Without an ozone layer to screen out some of this radiation, life could not have moved onto land.

Rise of the Eukaryotes

Nuclei are not often preserved during fossilization, but other traits provide evidence that a fossilized

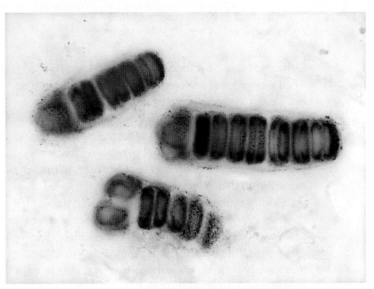

FIGURE 19.8 Fossil red algae (*Bangiomorpha pubescens*) that lived 1.2 billion years ago. Protists such as these algae were the earliest eukaryotes.

cell was eukaryotic. Eukaryotic cells are generally larger than prokaryotic ones. A cell wall with complex patterns, spines, or spikes probably belonged to a eukaryote. Researchers also look for biomarkers specific to eukaryotes. A **biomarker** is a substance that occurs only or predominantly in cells of a specific type. For example, certain steroids are found only in eukaryotes, so traces of these steroids are biomarkers for this group.

Steroids found in ancient rocks suggest eukaryotes may have arisen as early as 2.7 billion years ago. However, the oldest microfossils that most scientists agree are fossil eukaryotes date to about 1.8 billion years ago. The first eukaryotes were protists, and the oldest eukaryotic fossils that we can assign to a modern group are a type of red algae (**FIGURE 19.8**).

---

biomarker Molecule produced only by a specific type of cell; its presence indicates the presence of that cell.
stromatolite Dome-shaped structure composed of layers of bacterial cells, their secretions, and sediments.

TAKE-HOME MESSAGE 19.5

What was early life like and how did it change Earth?

✔ The cellular ancestor of all modern life arose by 3–4 billion years ago.

✔ The first cells were most likely anaerobic prokaryotes.

✔ By 2.5 billion years ago, oxygen released by photosynthetic cyanobacteria had begun to accumulate. The rise in oxygen concentration resulted in formation of the protective ozone layer and favored organisms capable of carrying out aerobic respiration.

✔ The first eukaryotic organisms may have arisen as early as 2.7 billion years ago. They were protists.

✔ Eukaryotic cells have a composite ancestry, with different components derived from different lineages.

Origin of the Nucleus

The DNA of most prokaryotes lies unenclosed in the cell's cytoplasm. By contrast, the DNA of a eukaryotic cell is always enclosed within a nucleus that is associated with an endomembrane system. The nucleus and endomembrane system probably evolved when the plasma membrane of an ancestral prokaryote folded inward (**FIGURE 19.9**).

Studies of the few types of bacteria that have internal membranes illustrates that such infolding can occur and how it can be advantageous. For example, some modern marine bacteria have membrane infoldings that greatly increase the surface area available to hold membrane-associated enzymes.

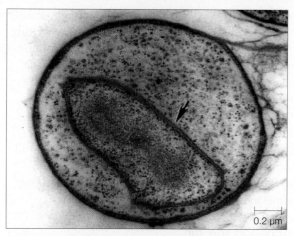

FIGURE 19.10 Nucleus-like structure in a prokaryote. DNA of *Gemmata obscuriglobus*, a species of bacteria, is enclosed by a double lipid bilayer membrane (indicated by the arrow).

Internal membranes also protect a genome from physical or biological threats. Consider *Gemmata obscuriglobus*, a bacterial species that withstands high levels of mutation-causing radiation. It is one of the few bacteria that has a membrane around its DNA (**FIGURE 19.10**). Like a eukaryotic nuclear envelope (Section 4.6), this membrane consists of two lipid bilayers. However, the membrane lacks nuclear pores or their equivalent, so it is not a true nuclear envelope. Researchers attribute the radiation resistance of *G. obscuriglobus* to the tighter packing, and therefore higher shielding, of its DNA within the membrane-enclosed compartment. DNA of other bacteria is more vulnerable to radiation because it is more spread out.

Origin of Mitochondria and Chloroplasts

Mitochondria and chloroplasts resemble bacteria in their size and shape, and they replicate independently of the cell that holds them. Like bacteria, they have their own DNA in the form of a single circular chromosome. They also have at least two outer membranes, with the innermost membrane structurally similar to a bacterial plasma membrane.

Recognition of these similarities led to the formulation of the **endosymbiont hypothesis**, which states that mitochondria and chloroplasts evolved from bacteria. (Endosymbiosis means "living inside" and refers to a relationship in which one organism lives inside another.) Nearly all eukaryotic lineages have mitochondria or mitochondria-like organelles, but only some have chloroplasts. Thus, biologists postulate that the two types of organelles were acquired independently in the sequence illustrated in **FIGURE 19.9**. In both

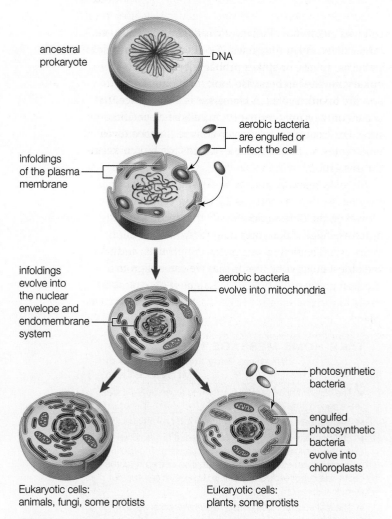

ancestral prokaryote — DNA

aerobic bacteria are engulfed or infect the cell

infoldings of the plasma membrane

infoldings evolve into the nuclear envelope and endomembrane system

aerobic bacteria evolve into mitochondria

photosynthetic bacteria

engulfed photosynthetic bacteria evolve into chloroplasts

Eukaryotic cells: animals, fungi, some protists

Eukaryotic cells: plants, some protists

FIGURE 19.9 ▶Animated Proposed steps in the evolution of some eukaryotic organelles.

CREDITS: (9) From Russell/Wolfe/Hertz/Starr, *Biology*, 2e, © 2011 Cengage Learning Inc.; (10) Courtesy of John Fuerst, University of Queensland. Originally published in *Archives of Microbiology* vol 175, p 413–29 (Lindsay MR, Webb RI, Strous M, Jetten MS, Butler MK, Forde RJ, Fuerst JA. Cell compartmentalisation in planctomycetes: Novel types of structural organization for the bacterial cell. *Arch. Microbiol.* 2001 Jun, 175(6): 413–29).

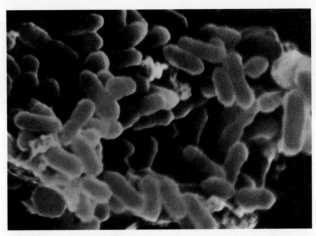

FIGURE 19.11 *Rickettsia prowazekii*, a bacterial species thought to be a close relative of the bacteria ancestral to mitochondria. It causes the disease typhus in humans.

cases, the process began when bacteria were taken up by or invaded a host cell, then replicated inside it. When the host cell divided, it passed some "guest" cells, referred to as endosymbionts, along to its offspring. As the two species lived together over many generations, genes carried by both partners were free to mutate. A gene could lose its function in one partner if the duplicate gene carried by the other partner still worked. Eventually, the host and endosymbiont lost enough duplicated genes to become incapable of living independently. At that point, the endosymbionts had evolved into organelles.

Given the evidence that mitochondria and chloroplasts evolved from bacteria, scientists are now investigating which modern bacteria are the closest relatives of these organelles. In such investigations, metabolic and genetic similarities between organelles and specific bacterial groups are considered to be evidence of shared ancestry.

So far, two groups of bacteria have been identified as close relatives of mitochondria. One group is the rickettsias, which are tiny bacteria that invade eukaryotic cells and replicate inside them. *Rickettsia prowazekii*, which causes the disease typhus, has a genome very similar to that of some mitochondria (**FIGURE 19.11**). The other potential mitochondrial relative is a type of free-living bacteria that floats in the ocean's surface waters. These marine bacteria have a genome similar to both rickettsias and mitochondria. Scientists do not know exactly how the two modern bacterial

groups relate to one another and to mitochondria. By one hypothesis, all three—mitochondria, rickettsias, and the modern marine bacteria—descended from an ancient free-living marine species.

The ancestry of chloroplasts is better understood. Cyanobacteria are the only modern bacteria that carry out photosynthesis by the oxygen-producing pathway, as chloroplasts do. Thus, ancient cyanobacteria are thought to have given rise to chloroplasts.

The endosymbiont hypothesis assumes that cells can enter and live inside other cells. It also assumes that such a relationship can, over time, become essential to the partners. Studies of modern-day cell partnerships lend support to both assumptions. One such study was carried out by microbiologist Kwang Jeon. In 1966, Jeon was studying *Amoeba proteus*, a species of single-celled protist. By accident, his amoebas became infected by a rod-shaped bacterium. Most infected amoebas died right away. A few, however, survived and reproduced despite their infection. Intrigued, Jeon maintained these infected cultures to see what would happen. Five years later, the descendant amoebas were host to many bacterial cells, yet they seemed healthy. In fact, when these amoebas were treated with bacteria-killing drugs that usually do not harm amoebas, they died. Apparently the amoebas had come to require the bacteria for some life-sustaining function. Further investigations revealed that the amoebas had lost the ability to make an essential enzyme. They now depended on their bacterial partners to make that enzyme for them.

Eukaryotic Divergence

However they arose, early eukaryotic cells had a nucleus, endomembrane system, mitochondria, and—in certain lineages—chloroplasts. These cells were the first protists. Over time, their many descendants came to include the modern protist lineages, as well as plants, fungi, and animals. The next section provides a time frame for these pivotal evolutionary events.

TAKE-HOME MESSAGE 19.6
How might the nucleus and other eukaryotic organelles have evolved?

✔ A nucleus and other organelles are defining features of eukaryotic cells.

✔ The nucleus and endomembrane system could have arisen through modification of infoldings of the plasma membrane.

✔ Mitochondria and chloroplasts most likely descended from bacteria.

---

endosymbiont hypothesis Hypothesis that mitochondria and chloroplasts evolved from bacteria.

CREDIT: (11) CNRI/Science Source.

19.7 Time Line for Life's Origin and Evolution

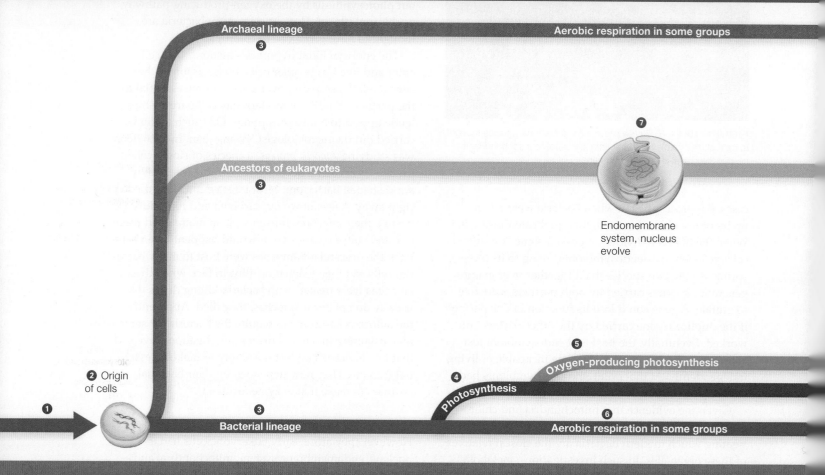

Hydrogen-rich, oxygen-poor atmosphere

Atmospheric oxygen level begins to increase

Archaeal lineage ❸

Aerobic respiration in some groups ❸

Ancestors of eukaryotes ❸

❼ Endomembrane system, nucleus evolve

❺ Oxygen-producing photosynthesis

❹ Photosynthesis

❷ Origin of cells

❶

❸ Bacterial lineage

❻ Aerobic respiration in some groups

3.8 billion years ago

3.2 billion years ago

2.7 billion years ago

Building Blocks Form

❶ Lipids, proteins, nucleic acids, and complex carbohydrates formed from the simple organic compounds present on early Earth.

Origin of Cells

❷ The first cells did not have a nucleus or other organelles. Oxygen was scarce, so the first cells made ATP by anaerobic pathways.

Domains Diverge

❸ An early divergence separated bacteria from the common ancestor of archaeal and eukaryotic cells. Not long after that, archaeal and eukaryotic cells diverged.

Photosynthesis and Aerobic Respiration Evolve

❹ Photosynthetic pathways that did not produce oxygen evolved in some bacterial lineages.

❺ Oxygen-producing photosynthesis evolved in a branch from this lineage, and oxygen began to accumulate.

❻ Aerobic respiration became the predominant metabolic pathway in some bacteria and archaea.

Endomembrane System and Nucleus Evolve

❼ Cell sizes and the amount of genetic information continued to expand in ancestors of what would become the eukaryotic cells. The endomembrane system, including the nuclear envelope, arose through the modification of cell membranes.

FIGURE 19.12 ▶Animated Milestones in the history of life, based on the most widely accepted hypotheses. As you read the next unit on life's past and present diversity, refer to this visual overview. It can serve as a simple reminder of the evolutionary connections among all groups of organisms. Time line not to scale.

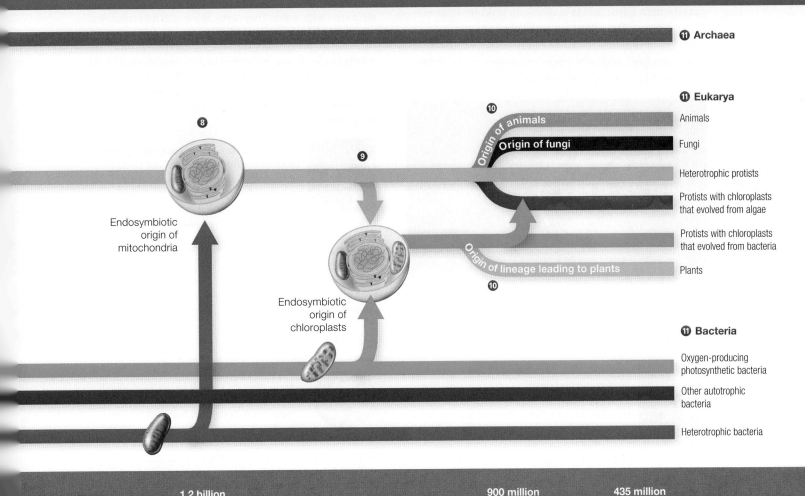

Atmospheric oxygen reaches current levels; ozone layer gradually forms

11 Archaea

11 Eukarya

Animals

Fungi

Heterotrophic protists

Protists with chloroplasts that evolved from algae

Protists with chloroplasts that evolved from bacteria

Plants

11 Bacteria

Oxygen-producing photosynthetic bacteria

Other autotrophic bacteria

Heterotrophic bacteria

8

Endosymbiotic origin of mitochondria

9

Endosymbiotic origin of chloroplasts

10 Origin of animals

Origin of fungi

10 Origin of lineage leading to plants

1.2 billion years ago

900 million years ago

435 million years ago

Origin of Mitochondria

8 Aerobic bacteria entered and lived inside a eukaryotic cell. Over many generations, descendants of these bacteria evolved into mitochondria.

Origin of Chloroplasts

9 An oxygen-producing, photosynthetic bacterial cell entered a eukaryotic cell. Over generations, bacterial descendants evolved into chloroplasts.

Origins of Fungi, Animals, and Plants

10 By 900 million years ago, representatives of all major lineages—including fungi, animals, and the algae that would give rise to plants—had evolved in the seas.

Modern Life

11 Modern organisms are related by descent, so all share certain traits. However, each lineage also has characteristic traits that evolved in response to the unique selective pressures it experienced.

FIGURE IT OUT From which lineage are mitochondria descended, bacteria or archaea?

Answer: Bacteria

Looking for Life (revisited)

When it comes to sustaining life, Earth is just the right size. If the planet were much smaller, it would not exert enough gravitational pull to keep atmospheric gases from drifting off into space. The photo below shows the relative sizes of Earth and Mars. As you can see, Mars is only about half the size of Earth. Mars has less gravity than Earth and is less able to hold on to atmospheric gases. The relatively small amount of atmosphere that remains consists mainly of carbon dioxide, some nitrogen, and only traces of oxygen. Thus, if life exists on Mars, it is almost certainly anaerobic.

summary

Section 19.1 **Astrobiology** is the study of life's origin and distribution in the universe. The presence of cells in deserts and deep below Earth's surface suggests life may exist in similar settings on other planets. Mars, our closest planetary neighbor, has some water, but lacks a protective **ozone layer**. Thus if any Martian life exists, it is likely deep in the soil.

Section 19.2 According to the **big bang theory**, the universe formed in an instant 13 to 15 billion years ago. Earth and other planets formed about 4.6 billion years ago from material released by explosions of giant stars. Early in Earth's existence, the planet had little or no free oxygen, and it received a constant hail of meteorites. Oxygen is highly reactive, so a lack of it would have facilitated assembly of simple organic compounds. Earth's surface was initially molten, but by 4.3 billion years ago it had cooled enough for life-sustaining water to pool on its surface.

Section 19.3 Laboratory simulations provide indirect evidence that organic monomers could have formed by lightning-fueled reactions in Earth's early atmosphere or in the hot, mineral-rich water around **hydrothermal vents**. Examination of meteorites shows that such compounds form in deep space and could have been transported to Earth by meteorites.

Section 19.4 Proteins that function in metabolic pathways might have first formed when amino acids stuck to clay, then bonded under the heat of the sun. The **iron–sulfur world hypothesis** postulates that metabolism began on the surface of rocks at hydrothermal vents.

Membrane-like structures and vesicles form when lipids are mixed with water. They serve as a model for **protocells**, which may have preceded cells.

An RNA world, an interval during which RNA was the genetic material, may have preceded the evolution of DNA-based systems. Discovery of **ribozymes**, RNAs that act as enzymes, lends support to the **RNA world hypothesis**. A subsequent switch from RNA to DNA would have made the genome more stable.

Section 19.5 Genetic comparisons among modern organisms indicate that all modern life is descended from the same cellular ancestor. The first cells were probably anaerobic and prokaryotic. Finding fossils of early cells is difficult. The oldest fossil evidence of cells is of photosynthetic bacteria that formed dome-shaped **stromatolites** and of bacteria that used sulfur as an electron acceptor. By 2.5 billion years ago, oxygen released as a by-product of photosynthesis by cyanobacteria had begun to change Earth's atmosphere. The increased oxygen level created a protective ozone layer, and favored cells that carried out aerobic respiration. **Biomarkers** and fossils of eukaryotes date back more than 2 billion years.

Sections 19.6, 19.7 Eukaryotes have a composite ancestry. Internal membranes of eukaryotic cells such as a nuclear membrane and endoplasmic reticulum probably evolved from infoldings of the plasma membrane. The **endosymbiont hypothesis** holds that mitochondria evolved from aerobic bacteria, and chloroplasts evolved from cyanobacteria.

Protists were the first eukaryotes. Over time, various protist lineages gave rise to plants, fungi, and animals. Evidence from many sources allows us to reconstruct the order of events and make a hypothetical time line for the history of life.

self-quiz   Answers in Appendix VII

1. According to the big bang theory, _____ .
   a. the universe expanded out from a single point
   b. Earth and our sun formed simultaneously
   c. carbon and oxygen were the first elements to form
   d. all of the above

2. An abundance of _____ in Earth's early atmosphere would have interfered with assembly of organic compounds.
   a. carbon dioxide   c. water
   b. ammonia   d. oxygen

data analysis activities

The Changing Earth Studies of ancient rocks and fossils can reveal changes that have taken place during Earth's existence. **FIGURE 19.13** shows how asteroid impacts and the composition of the atmosphere are thought to have changed over time. Use this figure and information in the chapter to answer the following questions.

1. Which occurred first, a decline in asteroid impacts, or a rise in the atmospheric level of oxygen (O_2)?

2. How do modern levels of carbon dioxide (CO_2) and O_2 compare to those at the time when the first cells arose?

3. Which is now more abundant, oxygen or carbon dioxide?

4. What do you think accounts for the rise in CO_2 at the far right of the graph?

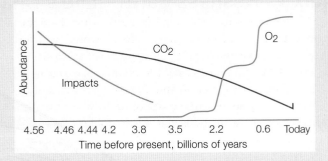

FIGURE 19.13 Changes in the abundance of asteroid impacts (green), atmospheric carbon dioxide concentration (pink), and atmospheric oxygen (blue) over geologic time.

3. Stanley Miller's experiment demonstrated that _____ .
   a. Earth is more than 4 billion years old
   b. under some conditions, amino acids can assemble spontaneously
   c. oxygen is necessary for all life
   d. DNA is less stable than RNA

4. According to one hypothesis, negatively charged clay particles played a role in early _____ .
   a. protein formation   c. photosynthesis
   b. DNA replication   d. oxygen declines

5. The prevalence of _____ in living organisms is taken as support for the idea that life arose near deep-sea vents.
   a. mitochondria   c. DNA
   b. iron–sulfide cofactors   d. a plasma membrane

6. An RNA that functions as an enzyme is a _____ .
   a. protein   c. ribosome
   b. protocell   d. ribozyme

7. Among prokaryotes, only the cyanobacteria _____ .
   a. live near hydrothermal vents
   b. produce oxygen during photosynthesis
   c. cannot tolerate oxygen
   d. have a nucleus-like structure

8. The evolution of _____ resulted in an increase in the levels of atmospheric oxygen.
   a. DNA-based genomes
   b. aerobic respiration
   c. sexual reproduction
   d. photosynthesis that releases oxygen

9. Bacteria that cause the disease typhus are close relatives of bacteria that evolved into _____ .
   a. protists   c. chloroplasts
   b. protocells   d. mitochondria

10. Infoldings of the plasma membrane into the cytoplasm of some ancestral cells may have evolved into the _____ .
   a. nuclear envelope   c. primary cell wall
   b. ER membranes   d. both a and b

11. An _____ is a relationship in which one organism lives inside another.

12. A stromatolite is a structure _____ .
   a. produced by endosymbiosis
   b. that formed only on the early Earth
   c. consisting of layered bacteria and sediment
   d. that expels hot water from deep in the Earth

13. The first eukaryotes were _____ .
   a. fungi   c. protists
   b. plants   d. animals

14. Evidence that Mars _____ suggests that it may have supported or still supports life.
   a. has an ozone layer
   b. has water
   c. is about the same size as Earth
   d. all of the above

15. Chronologically arrange the evolutionary events, with 1 being the earliest and 6 the most recent.
   ___ 1   a. onset of oxygen-releasing
   ___ 2   pathway of photosynthesis
   ___ 3   b. origin of mitochondria
   ___ 4   c. origin of protocells
   ___ 5   d. emergence of the first eukaryotes
   ___ 6   e. origin of chloroplasts
                f. the big bang

critical thinking

1. Researchers looking for fossils of the earliest life-forms face many hurdles. For example, few sedimentary rocks date back more than 3 billion years. Review what you learned about plate tectonics (Section 16.7). Then, explain why so few remaining samples of these early rocks remain.

2. The astronomer Carl Sagan once said, "We are made of star stuff." Explain why this is true of all life on Earth.

3. How did the evolution of oxygen-releasing photosynthesis in cyanobacteria increase the likelihood that mitochondria would one day evolve?

Actively-spreading ridges and transform faults

Total spreading rate, cm/year

Major active fault or fault zone; dashed where nature, location, or activity uncertain

Normal fault or rift; hachures on downthrown side

Reverse fault (overthrust, subduction zones); generalized; barbs on upthrown side

Volcanic centers active within the last one million years; generalized. Minor basaltic centers and seamounts omitted.

This NASA map summarizes the tectonic and volcanic activity of Earth during the past 1 million years. The reconstructions at far right indicate positions of Earth's major land masses through time.

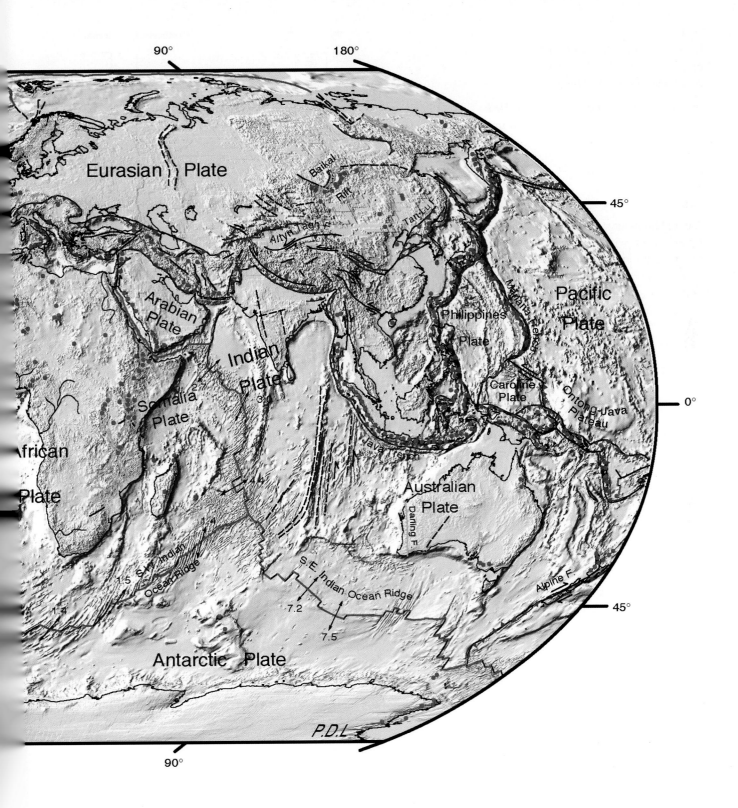

Appendix VI. Units of Measure

LENGTH

1 kilometer (km) = 0.62 miles (mi)
1 meter (m) = 39.37 inches (in)
1 centimeter (cm) = 0.39 inches

To convert	multiply by	to obtain
inches	2.25	centimeters
feet	30.48	centimeters
centimeters	0.39	inches
millimeters	0.039	inches

AREA

1 square kilometer = 0.386 square miles
1 square meter = 1.196 square yards
1 square centimeter = 0.155 square inches

VOLUME

1 cubic meter = 35.31 cubic feet
1 liter = 1.06 quarts
1 milliliter = 0.034 fluid ounces = 1/5 teaspoon

To convert	multiply by	to obtain
quarts	0.95	liters
fluid ounces	28.41	milliliters
liters	1.06	quarts
milliliters	0.03	fluid ounces

WEIGHT

1 metric ton (mt) = 2,205 pounds (lb) = 1.1 tons (t)
1 kilogram (kg) = 2.205 pounds (lb)
1 gram (g) = 0.035 ounces (oz)

To convert	multiply by	to obtain
pounds	0.454	kilograms
pounds	454	grams
ounces	28.35	grams
kilograms	2.205	pounds
grams	0.035	ounces

TEMPERATURE

Celsius (°C) to Fahrenheit (°F): $°F = 1.8 \ (°C) + 32$

Fahrenheit (°F) to Celsius: $°C = \dfrac{(°F - 32)}{1.8}$

	°C	°F
Water boils	100	212
Human body temperature	37	98.6
Water freezes	0	32

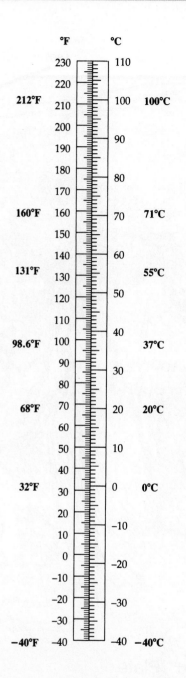

Appendix VII. Answers to Self-Quizzes

CHAPTER 16

1. b	16.2	
2. c	16.2	
3. d	16.3, 16.4	
4. b	16.4	
5. b	16.5	
6. e	16.5	
7. d	16.6	
8. Gondwana	16.7	
9. 66	16.8	
10. a	16.8	
11. a	16.4	
d	16.2, 16.5	
e	16.4	
f	16.6	
c	16.3	
b	16.3	
12. a	16.2–16.4	
c	16.7	
d	16.7	
e	16.5	
f	16.1	

CHAPTER 17

1. a	17.2	
2. d	17.2, 17.8	
3. a	17.6	
b	17.6	
4. d	17.7	
5. c	17.7	
6. b	17.8	
7. f	17.2, 17.3, 17.8	
8. a	17.9	
9. d	17.7, 17.9	
10. c	17.11	
11. a	17.10	
12. c	17.8	
d	17.7	
e	17.2	
b	17.8	
a	17.12	
f	17.12	
13. d	17.2, 17.12	

CHAPTER 18

1. c	18.2	
2. c	18.2	
3. d	18.2	
4. c	18.2	
5. b	18.2	
6. d	18.3	
7. d	18.3	
8. c	18.4	
9. b	18.4	
10. b	18.4	
11. True	18.4	
12. b	18.5	
13. e	18.2–18.5	
14. True	18.6	
15. b	18.2	
g	18.2	
d	18.5	
c	18.3	
e	18.4	
f	18.3	
a	18.2	

CHAPTER 19

1. a	19.2	
2. d	19.2	
3. b	19.3	
4. a	19.4	
5. b	19.4	
6. d	19.4	
7. b	19.5	
8. d	19.5	
9. d	19.6	
10. d	19.6	
11. endosymbiosis	19.6	
12. c	19.5	
13. c	19.5	
14. b	19.1	
15. (1) f	19.2	
(2) c	19.4	
(3) a	19.5	
(4) b	19.6	
(5) e	19.6	
(6) d	19.6	

Glossary of Biological Terms

adaptation (adaptive trait) A form of a heritable trait that enhances an individual's fitness in a particular environment. **256**

adaptive radiation Macroevolutionary pattern in which a burst of genetic divergences from a lineage gives rise to many new species. **288**

adaptive trait *See* adaptation.

allele frequency Abundance of a particular allele among members of a population, expressed as a fraction of the total number of alleles. **271**

allopatric speciation Speciation pattern in which a physical barrier arises and ends gene flow between populations. **284**

analogous structures Similar body structures that evolved independently in different lineages (by morphological convergence). **297**

astrobiology The scientific study of life's origin and distribution in the universe. **305**

balanced polymorphism Maintenance of two or more alleles of a gene at high frequency in a population. **278**

big bang theory Well-supported hypothesis that the universe originated by a nearly instant distribution of matter through space. **306**

biogeography Study of patterns in the geographic distribution of species and communities. **252**

biomarker Molecule produced only by a specific type of cell; its presence indicates the presence of that cell type. **311**

bottleneck Reduction in population size so severe that it reduces genetic diversity. **280**

catastrophism Now-abandoned hypothesis that catastrophic geologic forces unlike those of the present day shaped Earth's surface. **254**

character Quantifiable, heritable characteristic or trait. **294**

clade A group whose members share one or more defining derived traits. **294**

cladistics Making hypotheses about evolutionary relationships among clades. **295**

cladogram Evolutionary tree diagram that summarizes hypothesized relationships among a group of clades. **295**

coevolution The joint evolution of two closely interacting species; macroevolutionary pattern in which each species is a selective agent for traits of the other. **288**

comparative morphology The scientific study of similarities and differences in body plans. **253**

derived trait A novel trait present in a clade but not in any of the clade's ancestors. **294**

directional selection Mode of natural selection in which phenotypes at one end of a range of variation are favored. **274**

disruptive selection Mode of natural selection in which traits at the extremes of a range of variation are adaptive, and intermediate forms are not. **277**

endosymbiont hypothesis Hypothesis that mitochondria and chloroplasts evolved from bacteria that entered and lived inside another cell. **312**

evolution Change in a line of descent. **254**

evolutionary tree Diagram showing evolutionary connections. **295**

exaptation Evolutionary adaptation of an existing structure for a completely new purpose. **288**

extinct Refers to a species that no longer has living members. **288**

fitness Degree of adaptation to an environment, as measured by an individual's relative genetic contribution to future generations. **256**

fixed Refers to an allele for which all members of a population are homozygous. **280**

fossil Physical evidence of an organism that lived in the ancient past. **253**

founder effect After a small group of individuals found a new population, allele frequencies in the new population differ from those in the original population. **280**

frequency-dependent selection Natural selection in which a trait's adaptive value depends on its frequency in a population. **279**

gene flow The movement of alleles into and out of a population. **281**

gene pool All the alleles of all the genes in a population; a pool of genetic resources. **271**

genetic drift Change in allele frequency resulting from chance alone. **280**

genetic equilibrium Theoretical state in which an allele's frequency never changes in a population's gene pool. **272**

geologic time scale Chronology of Earth's history; correlates geologic and evolutionary events. **264**

Gondwana Supercontinent that existed before Pangea, more than 500 million years ago. **263**

half-life Characteristic time it takes for half of a quantity of a radioisotope to decay. **260**

homologous structures Body structures that are similar in different lineages because they evolved in a common ancestor. **296**

hydrothermal vent Underwater opening from which mineral-rich water heated by geothermal energy streams out. **307**

inbreeding Mating among close relatives. **281**

iron–sulfur world hypothesis Hypothesis that the metabolic reactions that led to the first cells took place on the porous surface of iron–sulfide-rich rocks at hydrothermal vents. **308**

key innovation An evolutionary adaptation that gives its bearer the opportunity to exploit a particular environment much more efficiently or in a new way. **288**

lethal mutation Mutation that alters phenotype so drastically that it causes death. **270**

lineage Line of descent. **254**

macroevolution Large-scale evolutionary patterns and trends; e.g., adaptive radiation, mass extinction. **288**

microevolution Change in an allele's frequency in a population. **271**

molecular clock Technique that uses molecular change to estimate how long ago two lineages diverged. **298**

monophyletic group An ancestor in which a derived trait evolved, together with all of its descendants. **294**

morphological convergence Evolutionary pattern in which similar body parts (analogous structures) evolve independently in different lineages. **297**

morphological divergence Evolutionary pattern in which a body part of an ancestor changes in its descendants. **296**

natural selection Differential survival and reproduction of individuals of a population based on differences in shared, heritable traits. Driven by environmental pressures. **256**

neutral mutation A mutation that has no effect on survival or reproduction. **270**

ozone layer High atmospheric layer rich in ozone; prevents most ultraviolet radiation in sunlight from reaching Earth's surface. **305**

Pangea Supercontinent that formed about 270 million years ago. **262**

parapatric speciation Speciation pattern in which two populations speciate while in contact along a common border. **287**

phylogeny Evolutionary history of a species or group of species. **294**

plate tectonics theory Theory that Earth's outer layer of rock is cracked into plates, the slow movement of which conveys continents to new locations over geologic time. **262**

Glossary of Biological Terms (continued)

population A group of organisms of the same species who live in a specific location and breed with one another more often than they breed with members of other populations. **270**

protocell Membranous sac that contains interacting organic molecules; hypothesized to have formed prior to the earliest life-forms. **309**

radiometric dating Method of estimating the age of a rock or a fossil by measuring the content and proportions of a radioisotope and its daughter elements. **260**

reproductive isolation The end of gene flow between populations. **282**

ribozyme RNA that functions as an enzyme. **308**

RNA world hypothesis Hypothesis that RNA served as the genetic information of early life. **308**

sexual dimorphism Difference in appearance between males and females of a species. **278**

sexual selection Mode of natural selection in which some individuals outreproduce others of a population because they are better at securing mates. **278**

sister groups The two lineages that emerge from a node on a cladogram. **295**

speciation Evolutionary process in which new species arise. **282**

stabilizing selection Mode of natural selection in which an intermediate form of a trait is adaptive, and extreme forms are not. **276**

stasis Evolutionary pattern in which a lineage changes very little over long spans of time. **288**

sympatric speciation Speciation pattern in which speciation occurs within a population, in the absence of a physical barrier to gene flow. **286**

theory of uniformity Idea that gradual repetitive processes occurring over long time spans shaped Earth's surface. **255**

Index

Index (continued)